MATERIAL AND MANUFACTURING TECHNOLOGY SERIES

ARTIFICIAL INTELLIGENCE IN MANUFACTURING RESEARCH

MATERIAL AND MANUFACTURING TECHNOLOGY SERIES

J. Paulo Davim, Editor

Tribology Research Advances
J. Paulo Davim
2009. ISBN: 978-1-60692-885-1

Drilling of Composite Materials
J. Paulo Davim (Editor)
2009. ISBN: 978-1-60741-163-5

Drilling of Composite Materials
J. Paulo Davim (Editor)
2009. ISBN: 978-1-60876-584-3 (Online book)

Artificial Intelligence in Manufacturing Research
J. Paulo Davim (Editor)
2010. ISBN: 978-1-60876-214-9

MATERIAL AND MANUFACTURING TECHNOLOGY SERIES

ARTIFICIAL INTELLIGENCE IN MANUFACTURING RESEARCH

J. PAULO DAVIM
EDITOR

NOVINKA
New York

LIBRARY OF CONGRESS CATALOGING-IN-PUBLICATION DATA

Artificial intelligence in manufacturing research / J. Paulo Davim, editor.
vi, 182 p. : ill. (some col.) ; 27 cm.
Includes bibliographical references and index.
ISBN 978-1-60876-214-9 (hardcover)
ISBN 978-1-62618-557-9 (softcover)
1. Manufacturing processes --Automation. 2. Manufacturing processes --Research. 3. Computer integrated manufacturing systems. 4. Artificial intelligence. I. Davim, Paulo J.
TS183 .A78 2010
670.285/63

2009031035

Published by Nova Science Publishers, Inc. † New York

Contents

PREFACE

Artificial intelligence is a subfield of computer science concerned with understanding the nature of intelligence and constructing computer systems capable of intelligent action. Artificial intelligence can be applied to all systems and manufacturing processes.

This book aims to provide the research and review studies on artificial intelligence in manufacturing. The first chapter provide information on application of neural networks and fuzzy sets to machining and metal forming. Chapter 2 discuss multi-objective optimization of multi-pass milling process parameters using artificial bee colony algorithm. Chapter 3 is focused on optimization of abrasive flow machining process parameters using particle swarm optimization and simulated annealing algorithms. Chapter 4 discuss the effects of process parameters on burr height in drilling of AISI 316 stainless steel using artificial neural network model. Chapter 5 is focused on artificial neural network modeling of surface quality characteristics in abrasive water jet machining of trip steel sheet. Subsequently, the chapter 6 deal with multi-objective optimization of cutting parameters for drilling aluminium AA1050. The chapter 7 is dedicated application of fuzzy logic in manufacturing, a study on modeling of cutting force in turning GFRP composites. The chapter 8 is dedicated on flank wear detection with AE signal and FNN during turning of Al/15vol%SiC-MMC. Finally, the last chapter of this research book is focused on integration of product development process using STEP and PDM.

The present research book can be used as a text book for final undergraduate engineering course (for example, mechanical, manufacturing, systems, etc) or as a subject on artificial intelligence in manufacturing at the postgraduate level. Also, this book can serve as a useful reference for academics, manufacturing and computational sciences researchers, mechanical, systems and manufacturing engineers, professional in related industries with artificial intelligence and manufacturing.

The Editor acknowledge their gratitude to Nova Publishers for this opportunity and for their professional support. Finally, I would like to thank all the chapter authors for their availability for this work.

J. Paulo Davim
Aveiro, Portugal
June 2009

RESEARCH AND REVIEW STUDIES

In: Artificial Intelligence in Manufacturing Research
Editor: J. Paulo Davim
ISBN 978-1-62618-557-9

Chapter 1

APPLICATION OF NEURAL NETWORKS AND FUZZY SETS TO MACHINING AND METAL FORMING

*U. S. Dixit**
Department of Mechanical Engineering,
Indian Institute of Technology Guwahati, India

ABSTRACT

Machining and metal forming are two major manufacturing processes. Neural networks and fuzzy sets have been employed in the modeling, optimization and control of these processes. This chapter describes how these soft computing tools can be used in the areas of machining and metal forming starting from the brief background of the methods. It is observed that neural networks are very effective tools of learning from the data. Multilayer perceptron and radial basis functions neural networks have been extensively used for this purpose. The advantages of fuzzy set based models are transparency of the model unlike black box nature of neural networks, ease of representing the experts' knowledge and ability to handle imprecise information.

In this chapter fuzzy set applications have been classified into three parts—application of fuzzy set operations, application of fuzzy arithmetic and application of fuzzy logic. The combination of fuzzy sets and neural networks is very effective and the techniques may be used in conjunction with finite element model. Some suggestions for future work have been provided.

1. INTRODUCTION

Of late neural networks and fuzzy sets are being applied in many areas of manufacturing. These techniques are the constituents of artificial intelligence, as they try to emulate human intelligence. In spite of development of a number of computational techniques and high speed computers, there is no substitute for skilled manufacturing personnel. It is, therefore, natural to focus the attention towards these techniques that resemble the thinking pattern of

---

* Corresponding author, E-mail: uday@iitg.ernet.in

experienced and skilled manufacturing personnel. Neural networks mimic the capability of a human to learn from the data and fuzzy sets mimic the ability to make inferences based on imprecise and linguistic information.

Fuzzy set and neural network based methods are parts of more general soft computing methods. Soft computing methods are distinguished from the conventional hard computing methods by the following characteristics:

- Soft computing methods can work with approximate data. Approximation occurs in the form of uncertainty and imprecision. Uncertainty refers to statistical variation and imprecision to the vagueness in the definition. As an example of uncertainty, consider that the flow stress of a material may be approximated by a normal distribution with a mean and a standard deviation. The more is the standard deviation, the more is the uncertainty in the flow stress value. As an example of imprecision, consider that the product A looks more beautiful than product B. One can not quantify the beauty in a crisp manner. Thus, there is an imprecision in the world 'beautiful'. Based on uncertain and imprecise data, soft computing methods can produce approximate but useful result. Although more often the results produced by the soft computing techniques are not accurate, one will usually get an idea of the extent of approximation.
- Soft computing methods can work with missing data. It is possible that in the presence of missing data, soft computing model may be more approximate and/or may not be available for the entire domain. However, usually, one will get an idea about the extent of approximation or certainty factor of the result obtained from the model.
- Soft computing methods have ability to learn from the data. One need not have a generalized theory of the phenomenon. One analogy can be given for the learning behavior of soft computing methods. Suppose one wants to learn driving a car. One method can be to teach him/her the physics and engineering of car and prepare a list of instruction for him/her to follow. Instructions can be of the form: "If you want to increase the speed by 10% in top gear, you have to press the accelerator by 3 mm". Imagine how effective this method of learning is! Many a times, hard computing methods have similar complexity. Another way of learning the car driving that is akin to soft computing methods is as follows. Let the learner start driving the car. He/she takes some actions and observes the result; then takes further action based on the feedback. After trying for a number of times, he/she will learn the driving. Many soft computing based methods learn from the data in the similar manner. A number of pairs of input (action) and output (result) are presented to a soft computing tool to train it, after which it becomes possible to predict the output not only for the training data but for other cases also.
- Soft computing algorithms are simple and many a times quite general in nature, such that the algorithm developed for one problem may be easily modified to make it suitable for the other case. For example, the methods to train a driver and a cook will be quite different if done in a hard computing way. However, doing in a soft computing way requires the similar training method. Both the cook and the driver have to learn from experience.

The present chapter describes the application of fuzzy set theory and neural networks in metal forming and machining. The main focus is on the application of these two major constituents of the soft computing. Only a cursory description of the theory of these techniques is provided in this chapter. The reader is encouraged to obtain the background of these methods from other sources [1–4].

2. Neural Network Applications in Machining

Neural networks map a point in a multi-dimensional input space to a point in a multi-dimensional output space. Assuming x_1, x_2, x_3,..........., x_n are the input variables and o_1, o_2, ..., o_k are the output variables, the mapping by a neural network can be represented as

$$o_i = f_i\left(x_1, x_2, x_3, \ldots\ldots\ldots, x_n\right), \quad i = 1, 2, \ldots\ldots\ldots, k \tag{1}$$

Thus, any output is a function of input variables. A function f_i is constructed by the neural network based on the learning data. Usually, data will be supplied in the form of input-output pair. The learning procedure utilizing the output (target) values is called supervised learning procedure. Sometimes, the learning data may contain only inputs and no output values. Then, also the neural network can learn using unsupervised learning procedure. The majority of applications of neural network in machining make use of supervised learning procedure. In the following subsection, two common types of neural networks are described.

2.1. Multi-Layer Perceptron and Radial Basis Function Neural Networks

For simplicity consider the case of two inputs and one output. Figure 1 shows a multi-layer perceptron network architecture with two hidden layer. Each layer consists of a number of neurons, which are basically the processing units. In this type of network, the weighted outputs of the neurons of one layer become the inputs to the neurons of the next layer. Each neuron of the input layers receives the value of one variable and emits it as it is to be used by subsequent layer. Let u_{ij} be the weight associated with i-th input neuron and j-th neuron in the first hidden layer. For the sake of simplicity, assume that each hidden layer consists of 3 neurons, although there is no restriction on the number of neurons in the hidden layers. If the processing-function of the j-th neuron in the first hidden layer is denoted by d_j, then the outputs of the neurons of the hidden layers are given by

$$\begin{aligned}
&\text{output of neuron 1}=d_1\left(u_{11}x_1 + u_{21}x_2\right),\\
&\text{output of neuron 2}=d_2\left(u_{12}x_1 + u_{22}x_2\right),\\
&\text{output of neuron 3}=d_3\left(u_{13}x_1 + u_{23}x_2\right).
\end{aligned} \tag{1}$$

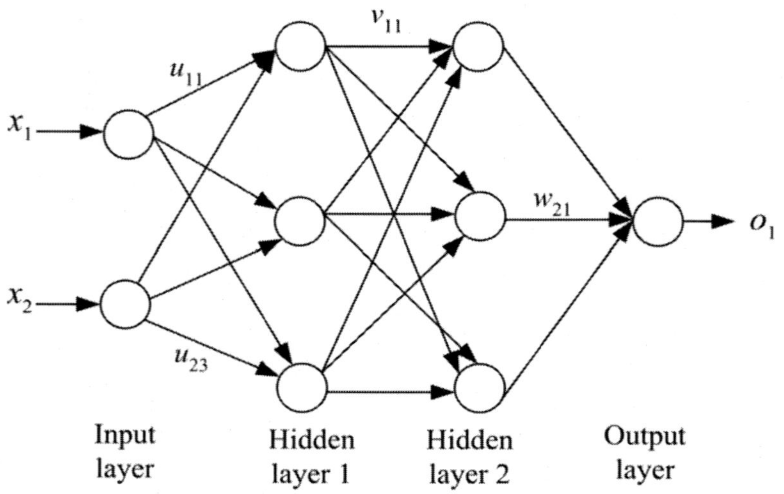

Figure 1. An MLP Neural Network.

The weighted outputs of hidden layer 1 reach hidden layer 2. Let v_{ij} be the weight associated with i-th neuron in the first hidden layer and j-th neuron in the second hidden layer. If the processing-function of the j-th neuron in the second hidden layer is denoted by e_j, then the outputs of the neurons of the hidden layers are given by

$$\begin{aligned}&\text{output of neuron 1}=e_1\left(v_{11}d_1+v_{21}d_2+v_{31}d_3\right),\\&\text{output of neuron 2}=e_2\left(v_{12}d_1+v_{22}d_2+v_{32}d_3\right),\\&\text{output of neuron 3}=e_3\left(v_{13}d_1+v_{23}d_2+v_{33}d_3\right).\end{aligned} \tag{2}$$

The weighted outputs of hidden layer 2 reach the output layer. Let w_{ij} be the weight associated with i-th neuron in the second hidden layer and j-th neuron in the output layer. If the processing-function of the j-th neuron in the second hidden layer is denoted by f_j, then the outputs of the neurons of the hidden layers are given by

$$o_1=f_1\left(w_{11}e_1+w_{21}e_2+w_{31}e_3\right). \tag{3}$$

In this architecture only two hidden layers are shown. In general, there can be more hidden layers.

Note that if all the processing functions are linear, the output will be a linear combination of input variables. However, if this were our aim, we would have used a multi-linear regression procedure. In general, output may be a non-linear function of the input variables. Hence, it is usual practice to have non-linear processing functions in the hidden layers. The function in the output layer is taken linear and often of the following form:

$$f(x)=x. \tag{4}$$

In that case, Equation (3) is written as

$$o_1 = w_{11}e_1 + w_{21}e_2 + w_{31}e_3. \tag{5}$$

It is clear that if all the processing functions are linear, the output will be a linear function of input. This is seldom the case. Usually, the processing functions of hidden layer neurons are nonlinear. By increasing the number of neurons in a hidden layer, it is possible to fit the function that matches the supplied data. Hence, many researchers find it convenient to have only one hidden layer.

In the above discussion, input to a neuron has been taken as the linear combination of the outputs of the neurons of the previous layer, with no constant term. A constant term or *bias* can be easily incorporated in it by having a neuron in each layer (except the output layer), which always emits 1. The weight associated with that neuron will act as bias.

The processing (activation) functions of the hidden layer neurons are non-linear function, which are bounded between 0 and 1 or between -1 and $+1$. One common function is the log sigmoid function given by

$$f(x) = \frac{1}{1+\exp(-cx)}, \tag{6}$$

which is bounded between 0 and 1 for $-\infty < x < \infty$. The parameter c is a slope parameter. Another common function is tan sigmoid given by

$$f(x) = \tanh(cx), \tag{7}$$

which is bounded between -1 and 1.

Having fixed the processing functions, the predictive capability of the neural network is decided by the weights. The procedure to obtain the proper weights is as follows:

- Divide total learning data into training, testing (called cross-validation by some authors) and validation (called testing by some authors) data. A typical ratio is 6:3:1.
- Determine the weights by any optimization algorithm, such that the error in prediction is minimized for training data upto certain error goal. With the trained neural network, the testing error is calculated. An optimum neural network provides reasonable predictions for training as well as the testing data. If the training error is less, but testing error is high, the network is called to over-fit the data. In that case, the network is trained again (usually with relaxed error goal) and tested.
- The network architecture and weights that provide the tolerable amount of error in training and testing data are frozen. The performance of the finalized neural network is assessed on the validation data.

The most common algorithm for the training of MLP neural networks has been backpropagation algorithm, in which the weights of the last layer are adjusted first and then the weights of other layers are adjusted in backward sequence. The optimization problem of determining the weights poses challenging issues. There is a tendency to get trapped into local minima. To avoid this problem, many non-traditional (particularly evolutionary) optimization algorithms have been employed. The convergence speed is also a major concern.

Another common neural network is radial basis function (RBF) neural network. It has only one hidden layer. A typical architecture consisting of 2 input, 3 hidden and one output neurons is shown in Figure 2. No weights are associated between hidden layer and input layer. The values of input variables reach the neurons of hidden layers as it is. With each neuron in the hidden layer is associated a center, which is a fixed vector, whose size is equal to the number of input variables. In fact, usually from the supplied learning data, some are chosen to act as center. The processing functions of the neurons in the hidden layer are radial basis functions, which map the Euclidean distance of an input vector from the center into a real number. The weighted outputs of each neuron in the hidden layer reach the output neuron, where it is summed. Thus, for the present neural network, the output is given by

$$o = \sum_{i=1}^{3} w_i \phi_i \left(\| \boldsymbol{x} - \boldsymbol{c}_i \|_2 \right), \tag{8}$$

where $\boldsymbol{x}$ is the input vector, $\boldsymbol{c}_i$ is the center of the i-th neuron in hidden layer, ϕ_i is the corresponding radial basis function and w_i is the corresponding weight. Specifically, for the network shown in Figure 2, the output is given by

$$o = \sum_{i=1}^{3} w_i \phi_i \left(\sqrt{(x_1 - c_{i1})^2 + (x_2 - c_{i2})^2} \right), \tag{9}$$

where c_{i1} and c_{i2} are the components of the center of i-th neuron. There are a number of possible radial basis functions. Some of them are as follows:

$$\text{Multiquadrics: } \phi_i(\boldsymbol{x}) = \left(\| \boldsymbol{x} - \boldsymbol{c}_i \|^2 + s_i^2 \right)^{1/2}, \tag{10}$$

$$\text{Inverse Multiquadrics: } \phi_i(\boldsymbol{x}) = \left(\| \boldsymbol{x} - \boldsymbol{c}_i \|^2 + s_i^2 \right)^{-1/2}, \tag{11}$$

$$\text{Gaussians: } \phi_i(\boldsymbol{x}) = \mathrm{e}^{-s_i^2 \| \boldsymbol{x} - \boldsymbol{c}_i \|^2} \tag{12}$$

$$\text{Thin Plate Splines: } \phi_i(\boldsymbol{x}) = \| \boldsymbol{x} - \boldsymbol{c}_i \|^2 \log \| \boldsymbol{x} - \boldsymbol{c}_i \|. \tag{13}$$

In Equations (10–12), s_i is called spread parameter that decides the zone of influence of a neuron. The most commonly used radial basis function for the modeling of metal forming and machining processes is Gaussian function.

From Equation (9) it is seen that once the centers and types of functions are fixed, the output is a linear function of the weights. The weights can be obtained by multiple-regression method. This procedure provides unique and fast solution.

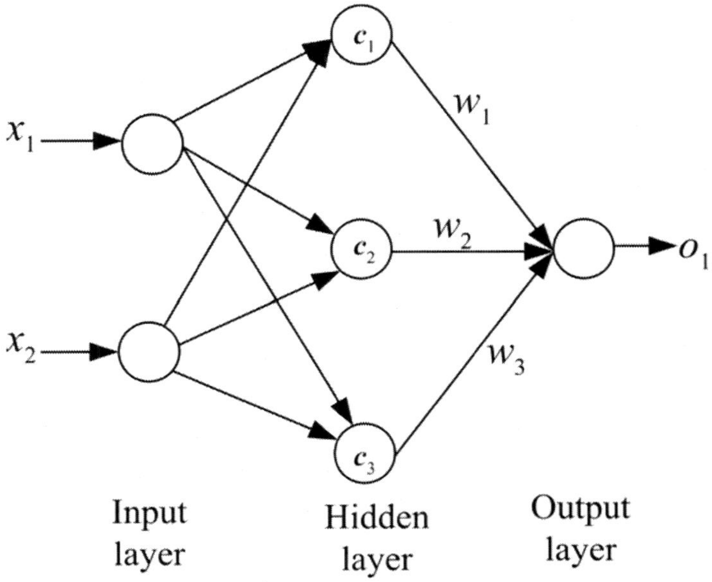

Figure 2. An RBF Neural Network.

2.2. Prediction of Machining Performance Using Neural Networks

The performance of a machining process is dependent on input variables and external factors. One can carry out the prediction online as well as offline. Online prediction requires the feedback from sensors. The important performance parameters that are modeled using neural networks are cutting force, surface roughness, dimensional deviation, temperature, tool life and tool wear. The sensors that are commonly employed for online prediction are accelerometer, cutting force dynamometer, acoustic emission sensor and sound measuring sensor. The network takes the cutting parameters like cutting speed, feed and depth of cut as input and predicts the required performance parameter as output.

Although there have been a number of attempts to model the machining performance parameters, there are a number of issues to be addressed. First challenge is to develop a neural network model that requires less amount of learning data. The second challenge is outliers. The experimental data often contains spurious data. Therefore, a data filtration scheme is required. The third challenge is the statistical variation in the data. A number of replicate experiments need to be carried out. The neural network model should be able to predict not only the parameters, but their expected variations. There have been few attempts to predict variations in the machining parameters. Kohli and Dixit [5] have used neural networks to predict lower, most likely and upper estimates of surface roughness in a turning process. Figure 3 shows their results for the case of dry turning of steel with coated carbide tools. It is seen that in almost all the cases, experimental values fall in between lower and upper estimates. There is a significant difference between upper and lower estimates indicating the presence of large amount of statistical variation.

Compared to the estimation of cutting force and surface roughness, estimation of tool wear is more difficult. Sick [6] reviewed 138 publications dealing with online and indirect tool monitoring in turning using neural networks and observed that a reliable tool wear monitoring system is still far away from the goal.

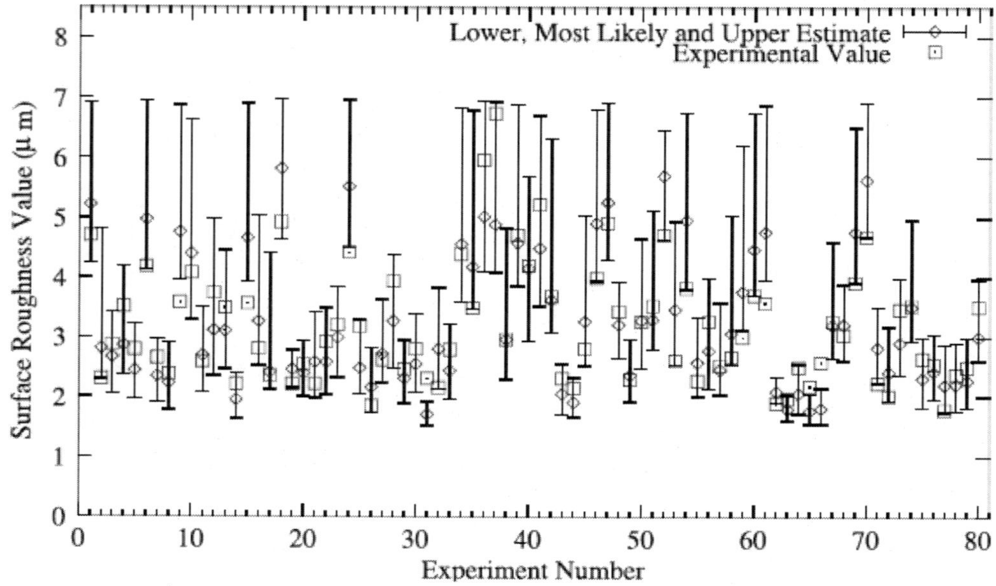

Figure 3. Predicted values versus experimental values of dry turning of steel with coated carbide tool. With permission from Kohli and Dixit [5]. Copyright [2005] Springer.

2.3. Optimization of Machining Performance Using Neural Networks

The optimization of machining process is an extensively researched area. The goals of optimization problems have been minimum cost of machining, maximum production rate and the weighted combinations of these two goals. These goals are subjected to various constraints such as maximum surface roughness, maximum cutting force, maximum power *etc.* Of late researchers have started solving multi-objective problems.

As a neural network model can provide performance of a machining operation as a function of input parameters, it can be used in the optimization process. The objective and constraint functions can be represented in the form of trained neural networks. Afterwards, any suitable optimization method can be employed for solving the optimization problem. Cus and Zuperl [7] have proposed a methodology of machining optimization with neural network. They optimize a manufacturer's implicit value function y that is a function of surface roughness, cost of machining and production rate. Thus, the three goals are combined into one depending on the manufacturer's preference. A neural network is trained with cutting speed, feed and depth of cut as input variables and manufacturer's value function y as the output variable. The training data is selected in such a manner, so that constraints are not violated. The trained neural network is used as the objective function and optimization is carried out by a suitable multi-variable optimization problem. Authors have provided an example of CNC turning, in which empirical expressions have been used for surface roughness, tool life, cutting force and cutting power. In place of empirical expressions, one can use neural networks for prediction of machining performance parameters.

Apart from offline optimization, online optimization of machining processes can also be carried out using neural networks. Ko and Cho [8] have proposed an approach for the adaptive optimization of face milling process that uses two neural networks— one for predicting the tool wear and the other for predicting the performance parameters (cutting force, cutting power and surface roughness). The first neural network used the feed force signal for the prediction of length of the wear. The input variables for the second neural network are cutting conditions and tool wear length. The process is optimized online using exterior penalty function method to tackle the constraints. Liu and Wang [9] have proposed neural network based control and optimization of milling process.

Neural network can be trained for producing the inverse relations too. One can develop a network that predicts surface roughness as a function of cutting speed, feed and depth of cut. Similarly, an inverse network can be developed to predict cutting speed, feed and depth of cut for a prescribed surface roughness. This will not be difficult if the inverse problem has a unique solution. One can suitably select the input variables and their ranges such that a unique solution is obtained. Using this concept, Azouzi and Guillot [10] has proposed an inverse process neurocontroller for optimizing the turning process. The force and vibrations sensors are used to provide the feedback in the proposed neurocontroller.

3. Neural Network Applications in Metal Forming

In metal forming area, neural networks have been used for the prediction of performance parameters and control. Most of the applications of neural networks are in the field of rolling, as the rolling is one of the oldest, the most important and the most complex process. Multi-layer perceptron network has found more application in the modeling and control of rolling process than any other type of network. Pican et al. [11] used MLP neural network with backpropagation algorithm for predicting the roll force in temper rolling. Temper rolling or skin-pass rolling is a cold rolling process, in which the sheet metal is subjected to very light reduction (0.5–4%) in thickness, in the presence of friction. The purpose of temper rolling process is to provide a degree of surface hardening, restore temper, prevent stretcher strains (Lueder's bands), impart a desired finish, and to impart a desired degree of flatness. The roll force in this process depends on a number of parameters. Pican et al. [11] have used the following parameters as the input parameters:

- Number of passes: 0, 1 or 2
- Width of sheet: 600–1900 mm
- Thickness of the sheet: 0.3–3.0 mm
- Percentage reduction: 0.2–2.0 %
- Lubrication: no lubrication, lubricant 1, lubricant 2
- Back tension: 10–50 N
- Front tension: 10–70 N
- Sheet temperature: 0–60 °C
- Surface finish of work rolls: 21 different conditions
- Roll wear (basically in terms of the total length of sheet processed by the work rolls): 0–130 km

- Diameter of work roll: 570–610 mm
- Flow stress (which is called as resistance by the authors): 120–430 MPa

The output layer just contains one neuron corresponding to roll force. The roll force prediction ultimately helps in adjusting the initial gap between the rolls. The final network contained two hidden layers— one with 48 neurons and other with 24 neurons. A total of 6221 training data and 1265 testing data were used for modeling. This data was collected from the plant in about six months. The requirement of huge amount of data for neural network modeling poses one of the challenges in its implementation.

Larkiola et al. [12] has predicted roll force by a combination of physics based model and neural network model. In their approach, although the roll force was calculated based on physics based model, the material and friction parameters in the model were obtained from trained neural networks. The neural network for the computation of material parameter (2 material constants) consisted of 18 input neurons corresponding to alloying elements, temperature of the last hot rolling pass, coiling temperature *etc* and was trained with 4500 training data. The final network contained one hidden layer with 10 neurons and provided root mean square error of 4.9%. The friction parameter was obtained in an inverse manner, so as to obtain the correct roll prediction by the physics based model. Thus, any error in the estimation of material parameter or physics based model is compensated by the friction parameter. The neural network for the computation of friction consisted of 10 input variables.

Cho et al. [13] used two different types of neural networks for the prediction of roll force in cold rolling. One MLP neural network predicted the roll force directly and other MLP neural network provided a corrective coefficient which has to be multiplied to the prediction made by a physics based model. The authors observed that both the MLP neural network models could improve the accuracy of prediction by 30–50% compared to the physics based model. Prediction by neural network is probabilistic in nature. Therefore, in order to have more reliability, the predictions of both neural networks and physics based model can be considered together. One can take the weighted average of three predictions. Lee and Lee [14] also proposed a model for the prediction of roll force in hot rolling that combines neural networks with the physics based model. Xie et al. [15] has used neural network for predicting the error in the mathematical model that predicts the coiling temperature. Recently, Geerdes et al. [16] applied the neural networks in the prediction of temperature in a hot strip mill. The authors observed that a combination of neural network model and physics based model provides the best result, in which the neural network is used to correct the physics based prediction.

Gorni [17] has described the application of neural networks in the modeling of hot plate rolling process for following tasks:

- Predicting the thermal profile of slabs in the reheating furnace
- Detecting turn-up of the plate, *i.e.*, detecting the excessive bowing up of the plate during rolling
- Predicting the length of the discarded portion of the final rolled stock, when the plate is rolled from cast slabs
- Pass scheduling *i.e.*, planning the reductions to be achieved in different stands of a tandem rolling mill.

Author has pointed out the requirement of sufficient amount of data for training as one of the challenges towards practical application of rolling process.

Gunasekera et al. [18] has developed a neural network model for cold rolling process. The data required in the modeling was collected from a physics-based model. The authors concentrated on reducing the error in prediction and training time of the neural network.

Among the physics based modeling techniques, finite element model (FEM) has been found to be very effective in predicting various performance parameters of rolling. However, FEM requires a large computational time and is not suitable for online prediction. A neural network can be trained with FEM generated data offline. The trained neural network can be used for predicting the performance parameter in subsequent applications, as the time required for prediction by a neural network is very small. This possibility has been explored by a number of researchers [19–21]. Among these, Dixit and Chandra [19] concentrated on reducing the number of training and testing data for neural network modeling and prediction the most likely, lower and upper estimates of roll force and roll torque.

Like in the rolling, neural networks can be employed to model other metal forming processes. However, there is less number of papers describing the application of neural networks in the other processes. Osakada and Yang [22] employed neural networks for the process planning of cold forging. Raj et al. [23] has applied neural networks in the modeling of hot extrusion and hot upsetting. Mori and Li [24] applied neural networks in the forging process. The authors used two geometric parameters of the die and aspect ratio of the workpiece as the input variables of the neural network model. The corresponding outputs were the die filling state and the forging load. The training and testing data for the neural network were obtained from FEM simulations. Kazan et al. [25] predicted spring-back in wipe bending using neural networks.

There have been far less number of applications of RBF neural networks for the prediction of performance parameters in rolling. Sbarbaro-Hofer [26] used RBF neural networks in the control of the strip thickness in a steel rolling mill. Recently, Gudur and Dixit [27] used RBF neural network for predicting the roll force and roll torque in cold rolling process. In their model, network had 5 input parameters. They used 55 training data and 30 testing data. The percentage root mean squared fractional errors in the roll force prediction were 0.25% for training and 2.9% for testing. The corresponding values for the roll torque were 0.77% and 3.5%.

The above description has highlighted the application of neural networks to metal forming in the following ways: (1) use of neural network as an alternative computational tool for the modeling of the process (2) use of neural network for correcting the predicted values obtained by physics-based methods and (3) capturing the knowledge of physics-based model with a view point of reducing the prediction time for online learning and/or optimization. Gudur and Dixit [28] have presented one more application of neural network— its use as an assistive tool to FEM model of the process. The general overview of the scheme for cold flat rolling is shown in Figure 4. The authors have used a plane strain rigid-plastic FEM model that provides the velocity components at each node alongwith the location of neutral point. A radial basis function neural network (RBF NN) accepts these as input and gets trained. The trained neural network is kept in a repository. Now, whenever a problem is to be solved, initial velocity field and neutral point are obtained from this trained neural network. Further iterations of FEM are carried out to obtain a more accurate solution and post-process the results. It is observed that the reduction in the computational time is by a factor of order 10.

For a more refined mesh updated Lagrangian scheme, the procedure is expected to provide more significant time saving.

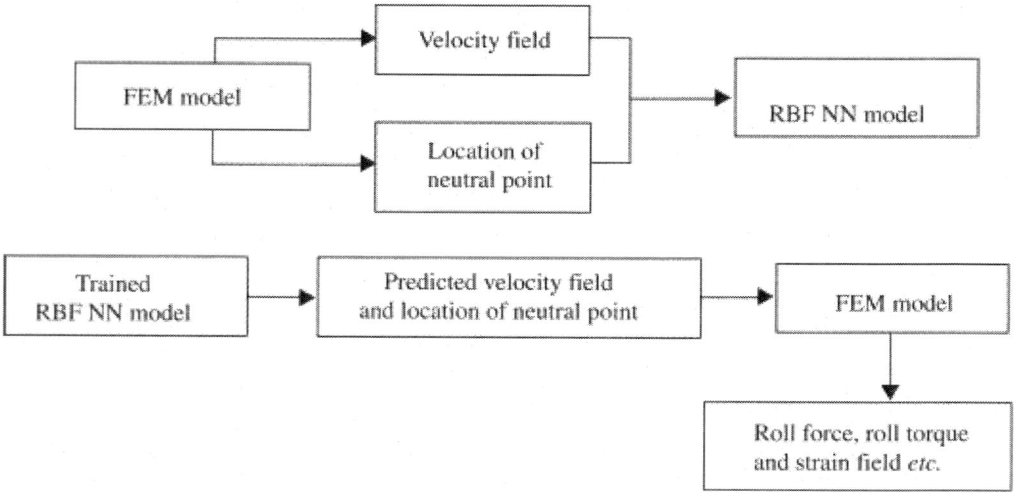

Figure 4. A general overview of neural network assisted FEM model. With permission from Gudur and Dixit [28]. (A label in the figure has been slightly modified). Copyright [2005] Elsevier.

4. Fuzzy Set Applications in Machining

A fuzzy set is the set in which its members are allowed to have any positive membership grade between 0 and 1. A membership grade denotes the strength of association of the element within the set with 1 indicating the perfect association. To understand the concept of fuzzy sets, consider the universe containing the first nine natural numbers from 1 to 9. Suppose that one has to construct a fuzzy set F of "numbers close to 5" from these numbers. There can be various solutions to this problem. Some of the solutions are as follows:

$$\begin{aligned} F &= \{0.2/3, 0.6/4,\ 1/5,\ 0.6/6,\ 0.2/7\}, \\ F &= \{0.3/3, 0.7/4,\ 1/5,\ 0.7/6,\ 0.3/7\}, \\ F &= \{0.1/2, 0.4/3, 0.9/4,\ 1/5,\ 0.9/6,\ 0.4/7, 0.1/8\}, \end{aligned} \quad (14)$$

where 0.2/3 means that the membership grade of natural number 3 is 0.2. Note that in all the above solutions, the natural number 5 is having the membership grade 1 as there is no doubt that it is close to 5. On the other hand, there is subjectivity in the allocation of membership grades to other numbers. This depends on how one perceives the word 'close'. To find out suitable values of membership grades is one of the most challenging tasks of fuzzy set theory. It is natural that it needs some art in capturing the human thinking in the mathematical framework.

Fuzzy sets have been extensively used in decision making by employing various operations of fuzzy set theory. As a simple example of it, consider three cutting fluids L1, L2 and L3. Let A be the set of high specific heat and B the set of high thermal conductivity and

$$A = \{0.7/\text{L1},\ 0.9/\text{L2},\ 0.8/\text{L3}\},\quad B = \{0.8/\text{L1},\ 0.5/\text{L2},\ 0.6/\text{L3}\} \tag{15}$$

As per expert opinion, a good cutting fluid should have high thermal conductivity and high specific heat. In fuzzy set theory, the word 'and' can be replaced by an intersection operator. Thus, the fuzzy set of high specific heat and high thermal conductivity is given by

$$A \cap B = \{0.7/\text{L1},\ 0.5/\text{L2},\ 0.6/\text{L3}\}. \tag{16}$$

Note that here the following definition of $A \cap B$ has been used. The membership grade of an element in the set $A \cap B$ is the minimum of its membership grades in A and B. Now, examining the set $A \cap B$ (Eq. 16), it is seen that L1 has the highest membership grade and thus is the most suitable cutting fluid. In the same way, the 'or' of English language can be replaced by a union operation of the fuzzy set, which is usually defined as follows. The membership grade of an element in set $A \cup B$ is the maximum of its membership in A and B.

A prominent branch of fuzzy set theory is the fuzzy arithmetic, which deals with the fuzzy numbers. Fuzzy numbers are the generalization of interval number, where corresponding to each membership grade, a different interval can be assigned. For example, coefficient of friction in a machining process may be (0.7, 0.7), (0.6, 0.8) and (0.5, 0.9) at membership grades of 1, 0.8 and 0.5 respectively. Many of the process variables are imprecise and uncertain and thus are best represented by fuzzy numbers. When these variables are put in a physics-based model, the result is obtained in the form of fuzzy numbers, which is a more realistic picture. Note that apart from using fuzzy variables in a physics model, the fuzziness can be introduced in the model itself, if the physics is uncertain. For example, assume a physics-based model provides that the parameter y is related to parameter x in the following way:

$$y = x^2. \tag{17}$$

This can be converted to fuzzy model by using the following relation:

$$y = x^a, \tag{18}$$

where a is a fuzzy number, whose most likely value corresponding to the membership grade 1 is 2. Once the result is obtained in the form of fuzzy numbers, using this information in an appropriate way is another important task.

Another important branch of fuzzy set theory is fuzzy logic, which uses IF-THEN rules in which either the antecedent (IF part) or the consequent (THEN part) or both contain fuzzy sets. The decision is taken by evaluating the fuzzy rules. In Mamdani fuzzy inference system, both antecedent and consequent are fuzzy sets. Sugeno fuzzy model, also known as Takagi, Sugeno and Kang (TSK) model contains fuzzy sets in antecedent and crisp function in the consequent. A typical rule in Mamdani system is as follows:

If depth of cut is low and feed is low and cutting speed is high then surface roughness is low.

The same rule in TSK system may look as follows:

If depth of cut is low and feed is low and cutting speed is high then surface roughness is $5 fv^{-0.1}\sqrt{d}$.

In the following paragraphs, some applications of fuzzy set theory to machining will be described. The applications have been classified into three groups— (i) Application of fuzzy set operations, (ii) Applications of fuzzy arithmetic, (iii) Applications of fuzzy logic.

4.1. Applications of Fuzzy Set Operations

Fuzzy set operations are very helpful in taking decisions in the presence of conflicting and incommensurable objectives. For example, consider that 5 different types of models of a product have to be evaluated against two objectives. If a model has a membership grade μ_1 in one objective and μ_2 in other objective, its overall membership grade can be defined as

$$\mu_o = \min(\mu_1, \mu_2). \tag{19}$$

Here, the overall performance of the model is dependent on the most poorly performing aspect. This type of strategy is called non-compensating strategy. An overall membership grade based on a compensating strategy may be defined

$$\mu_o = \sqrt{\mu_1 \mu_2}. \tag{20}$$

A weighted combination of two strategies may also be taken. Thus, if membership grades of a model in n different objectives are $\mu_1, \mu_2, \ldots\ldots\ldots, \mu_n$, then overall membership grade may be defined as

$$\mu_o = (1-\alpha)\min(\mu_1, \mu_2, ,,,,, \mu_n) + \alpha \sqrt[n]{\mu_1 \mu_2 ,,, \mu_n}, \tag{21}$$

where α is a weight factor. If α=1, then Eq. (21) provides a pure compensating strategy and if α=0, it provides a pure non-compensating strategy.

One can easily incorporate linguistic information in the decision process. For example, if a component has a membership grade μ in a set of 'good surface finish', it can be assigned a membership grade of $\sqrt{\mu}$ in the set of 'more or less good surface finish'. Thus, customers' requirement available in the language form may be converted into mathematical form.

In the area of multi-objective machining optimization, the fuzzy set based decision strategy may be very effective. There are a number of objectives to be satisfied in a multi-objective machining optimization such as minimum cost of product, maximum production rate, maximum profit rate and minimum surface roughness. In the fuzzy set based approach, different solutions may be assigned membership grades in different objectives and then based on the fuzzy set operations, the solution having the overall membership grade may be chosen. Panchal et al. [29] have used this type of approach in the optimization of turning process.

4.2. Applications of Fuzzy Arithmetic

Fuzzy arithmetic is helpful when the values of variables are uncertain and imprecise. Several researchers have used fuzzy arithmetic in machining optimization. Recently, Liu [30] has developed a method that can obtain the optimum fuzzy cost of machining in a multi-pass turning process, when the exponents and coefficients in the objective and constraint functions are taken as fuzzy number. The optimum cost in the form of a fuzzy number provides more information than the crisp value.

In developing a fuzzy arithmetic model, one may face difficulty in assigning suitable fuzzy numbers to various variables. A simple way is to treat these variables as linear triangular numbers. The most likely estimate of a variable can be assigned a membership grade 1 and the lower and upper estimates of the variable can be assigned membership grades of 0.5. With these three points a triangle can be constructed. The more sophisticated methods can also be tried.

4.3. Applications of Fuzzy Logic

The fuzzy logic containing the IF-THEN fuzzy rules has been used in developing the expert system and controller. Basically, a fuzzy logic based controller may be considered as an expert system that can take the appropriate decision for a control action based on the sensory feedback. Thus, software part of a controller is similar to software of an expert system. Fuzzy logic has been widely used in modeling and control of machining problems.

Figure 5 shows the block diagram of an expert system that can learn from shop floor data. The system provides the necessary output that can be used for either process planning or control. The expert system requires IF-THEN rules, which can either be supplied by an expert or can be generated from the shop floor data. The shop floor data is limited in quantity. It often becomes difficult to use that data for generating all the required rules. A good strategy can be to train a neural network from the data, as neural networks are efficient in learning from the data. The trained neural network can be used to generate lots of IF-THEN rule. The strategy described in [31, 32] can be adopted for generating the rules. The generated rules can be evaluated by an expert and expert can suitably modify them. Once the rule-base is available, the inference module can predict the output for given set of input.

Abburi and Dixit [31] used this procedure for the surface roughness prediction in a turning process. The input variables of their expert system are cutting speed v, feed f, depth of cut d and acceleration of radial vibration of the tool-holder a. Figure 6 shows how the input and output variables are fuzzified into a number of overlapping fuzyy subsets. In this type of fuzzification, the value of a particular variable may be part of two fuzzy subsets. Accordingly, for a given set of input variables, a number of rules get fired. The strength of different rules is different. The combined action of the rules decides the output of the expert system based on the standard procedure of fuzzy logic theory. The validation of the procedure with experimental data is good. The similar procedure may be employed for modeling of other machining parameters such as cutting force, tool wear, and tool life. The expert system can also be conveniently used for solving the inverse problems. For example, for a given surface roughness, cutting speed, feed and depth of cut can be obtained. This type of expert system also possesses a certain amount of extrapolation capability, unlike neural networks.

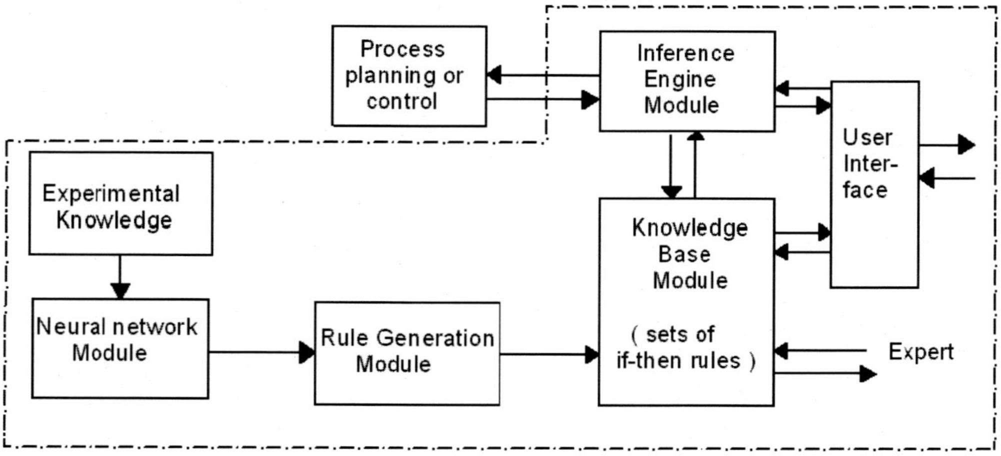

Figure 5. A block diagram of an expert system that has learning capability. With permission from Abburi and Dixit [31]. Copyright [2006] Elsevier.

The fuzzy set based expert system developed in [31] has the following advantages:

- The system is able to learn from exemplars.
- Based on the input parameters, the system can predict the output quickly.
- The system is transparent. The user can see the rules and find the flaw in the system, if any.
- The system can solve the inverse problems quickly. The inverse problems usually have multiple solutions, which the expert system can easily identify.
- A certain amount of extrapolation is possible, as the physics of the process usually do not change abruptly.

In an early work, Fang and Jawahir has [33] has proposed a fuzzy logic based methodology to assess total machining performance encompassing surface finish, tool-wear rate, dimensional accuracy, cutting power and chip breakability. The authors quantified the effect of major influencing factors on total machining performance by fuzzy set method and developed a series of fuzzy set models to give quantitative assessments for the given set of input conditions. Fang et al. [34] has presented a fuzzy set based methodology for predicting the chip form and chip breakability in machining. A series of machining experiments were conducted to establish a chip control database which is used as a standard to quantify the variability of chip breaking due to effects varying process parameters.

In References [35, 36], process parameters are selected/optimized based on fuzzy logic. Susanto et al. [37] applied fuzzy logic in tool wear monitoring of face milling process. References [38, 39] deal with predicting burns in grinding process. References [40–42] apply fuzzy logic to drilling process.

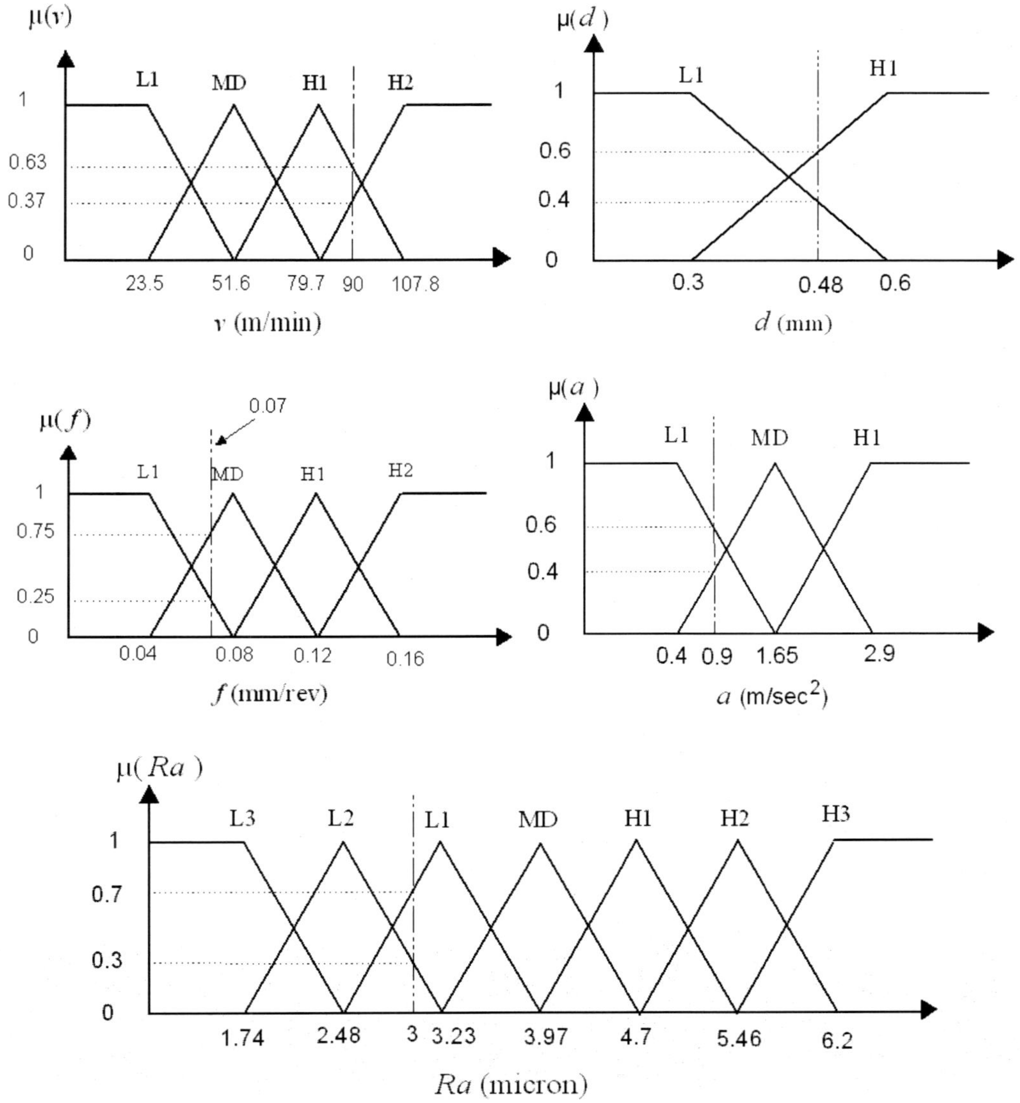

Figure 6. Fuzzification of input and output variables for turning process..With permission from Abburi and Dixit [31]. Copyright [2006] Elsevier.

5. Fuzzy Set Applications in Metal Forming

Fuzzy set theory has been applied in modeling and control of metal forming processes. Amongst metal forming processes, rolling is one of the most investigated processes. Therefore, there are a number of papers dealing with the application of fuzzy sets to it. In the following sub-sections, the application of fuzzy sets to rolling process is described. The fuzzy set theory can be applied to other metal forming processes in the similar way.

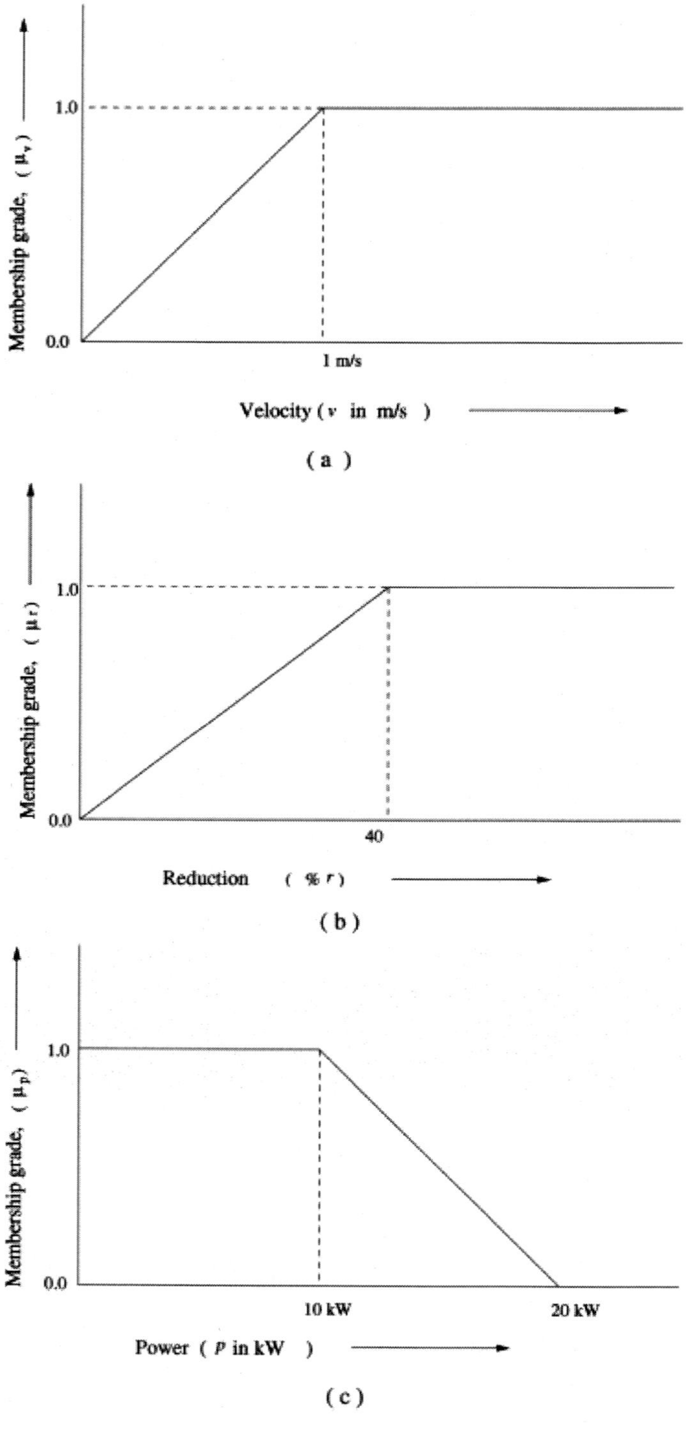

Figure 7. Membership functions of desired: (a) velocity (b) reduction and (c) power. With permission from Dixit et al. [43]. Copyright [2002] Elsevier.

5.1. Applications of Fuzzy Set Operations

An interesting application of fuzzy set operations has been provided in reference [43]. Here, the fuzzy set operations have been used for the conceptual design of a laboratory cold rolling mill. The customer's requirement of maximum mill speed, maximum possible reductions and rolling power are represented as fuzzy sets as shown in Figure 7.

The customer will be very pleased to have a mill speed of more than 1m/s as indicated by the constant membership grade of 1.0 after an outlet strip velocity of more than 1 m/s. As the strip-velocity reduces below 1 m/s, the membership grade indicating the satisfaction level of the customer linearly decreases to become 0 at 0 strip-velocity. In the same way, the membership functions of desired reduction and power may be interpreted. Note that Figure 7 (c) represents a fuzzy set of low power. The membership grade in the fuzzy set of very low power is obtained by squaring its membership grade in the fuzzy set of low power. Assuming that the customer requires high reduction, high mill-speed and very low power, the overall membership grade is obtained by the fuzzy intersection operation. Thus, the overall membership grade μ_0 is given by

$$\mu_o = \min\left(\mu_r, \mu_v, \mu_p^2\right). \tag{22}$$

Among the various possible solutions, the solution providing the maximum overall membership grade is chosen.

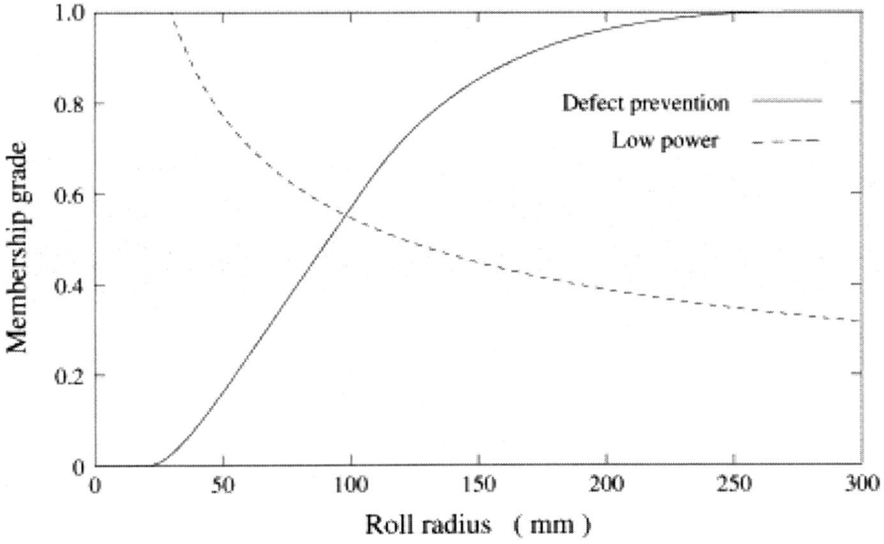

Figure 8. Membership functions of roll diameter from the viewpoints of low power and defect-free rolling. With permission from Dixit et al. [43]. Copyright [2002] Elsevier.

In reference [43], the fuzzy set theory has also been applied for deciding the roll diameter. Large diameter rolls provide better rigidity, better cooling, lesser value of the minimum coefficient of friction, a greater value of maximum possible reduction and lesser tendency of central bursting and split end defects in the rolled product. On the other hand, small rolls provide lesser rolling power and lesser spread. Thus, there are conflicting objectives in the selection of proper roll diameter. The proper roll diameter will maximize the overall objective, which is measured by the overall membership grade. For example, Figure 8 shows the membership functions of roll diameter from the viewpoints of low power and defect-free process. The intersection of two curves is the point that maximizes the overall membership grades in these two objectives. In the same way, other objectives may be considered. The art lies in choosing the proper shape functions.

Dixit and Dixit [44] applied fuzzy set operations in the scheduling of tandem mills. The conflicting objectives of minimum power and maximum reliability were optimized based on the fuzzy set theory. In this work, the material parameter and friction are considered as fuzzy number.

5.2. Applications of Fuzzy Arithmetic

Metal forming problems have been extensively modeled using finite element method. A finite element package requires the data about the material properties and friction coefficient. Often these values are imprecisely known and can therefore be represented as fuzzy numbers.

Dixit and Dixit (1996) considered the material and friction parameters as fuzzy number and obtained the values of roll force, roll torque and roll pressure distribution from a finite element model. The material is assumed to strain harden in the following way:

$$\sigma_y = \left(\sigma_y\right)_0 \left(1+\frac{\varepsilon_{eq}}{b}\right)^n, \tag{23}$$

where $\left(\sigma_y\right)_0$ is the yield stress of the raw material, ε_{eq} is the equivalent strain and b and n are material handling constants. The parameters $\left(\sigma_y\right)_0$, b and n are taken as fuzzy. Figure 9 shows the membership functions of $\left(\sigma_y\right)_0$, b and n for a steel and coefficient friction. Based on these parameters, variation of roll force with reduction for some membership grades is shown in Figure 10. Reference [11] mentioned in the figure pertains to the work of Shida and Awazuhara [46].

5.3. Applications of Fuzzy Logic

Fuzzy logic has been widely used for controlling the various processes including rolling. Recently, Gudur and Dixit [47] employed it for the modeling of roll force and roll torque. For this purpose, a Sugeno fuzzy model is used and the methodology of reference [48] is adopted. The representation of IF-THEN rules in the TS fuzzy model is given as

$$R_i : \text{IF } x_1 \text{ is } A^1(i) \text{ and } x_2 \text{ is } A^2(i) \ldots\ldots\text{and } x_k \text{ is } A^k(i), \\ \text{THEN } \; y_i = a_i^0 + a_i^1 x_1 + a_i^2 x_2 + \ldots\ldots + a_i^k x_k. \tag{24}$$

where R_i is the ith rule, vector $\boldsymbol{x} = [x_1, x_2, \ldots x_k]$ represents the set of k input variables like reduction, roll radius *etc* and $A^j(i)$ is the fuzzy subset corresponding to the jth input variable. Note that depending on the rule, $A^j(i)$ can be low (L), medium (MD) or high (H). The output of the ith IF-THEN rule is y_i, which is assumed to be a linear function of the input variables. The coefficients associated with the input variables in the linear functions are denoted as a_i^j, where subscript denotes the rule number and superscript denotes the associated input variable. The aggregated output of the Sugeno fuzzy model, $\hat{y}$ is expressed by a weighted average of the rule consequents:

$$\hat{y} = \frac{\sum_{i=1}^{N_R} w_i y_i}{\sum_{i=1}^{N_R} w_i}, \tag{25}$$

where w_i is the strength of the ith fuzzy rule and N_R is the total number of rules. The strength of the rule is taken as the product of the membership grades of the input values of the variables in the corresponding fuzzy subsets.

The coefficients a_i^j of Eq. (24) can be obtained by minimizing the maximum approximation error (λ) between the actual output and the fuzzy model output over the given input-output datasets, which can be obtained experimentally or from FEM model. For this, the following linear programming (LP) model is solved:

$$\begin{aligned} &\text{Minimize } \lambda, \\ &\text{subject to,} \\ &o_p - \hat{y}_p \le \lambda, \qquad p = 1, 2, \ldots\ldots m, \\ &-o_p + \hat{y}_p \le \lambda, \qquad p = 1, 2, \ldots\ldots m, \\ &\qquad\qquad\qquad \lambda \ge 0, \\ &a_i^j \ge 0, \qquad i = 1, 2, \ldots\ldots N_R, \quad j = 1, 2, \ldots k. \end{aligned} \tag{26}$$

where the subscript p represents the pth input-output pattern and m is the total number of patterns in the dataset. The actual output of the pth data is o_p. The maximum approximation error is termed as λ. The design variables of the problem are a_i^j and λ.

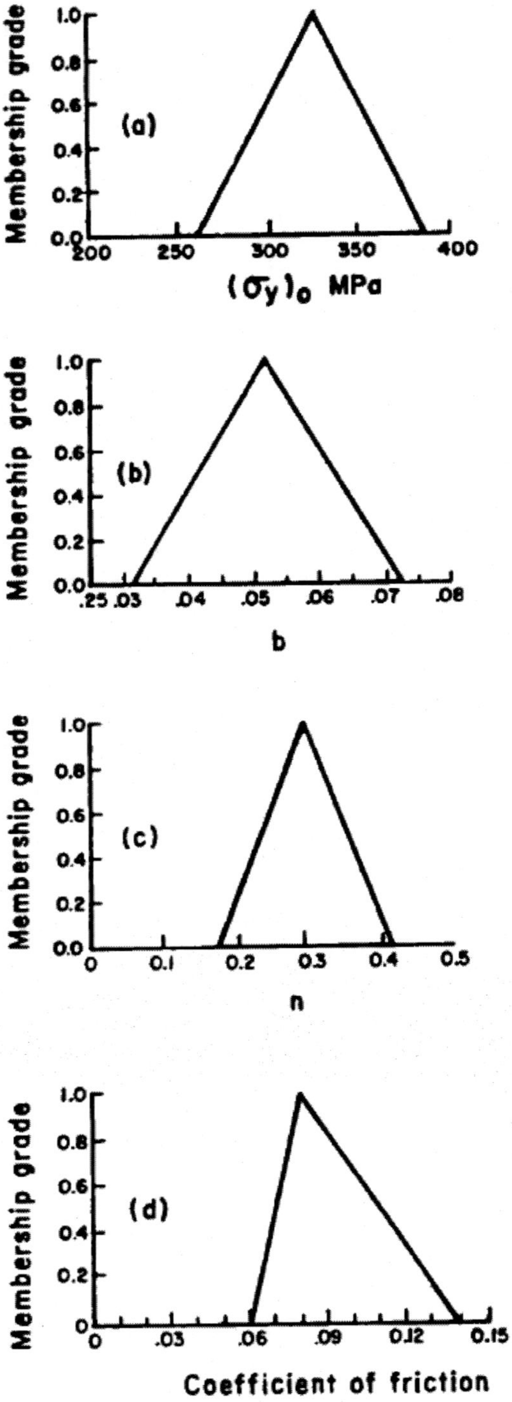

Figure 9. Membership functions of (a) $\left(\sigma_y \right)_0$ (b) b (c) n (d) coefficient of friction. With permission from Dixit and Dixit [45]. Copyright [1996] Elsevier.

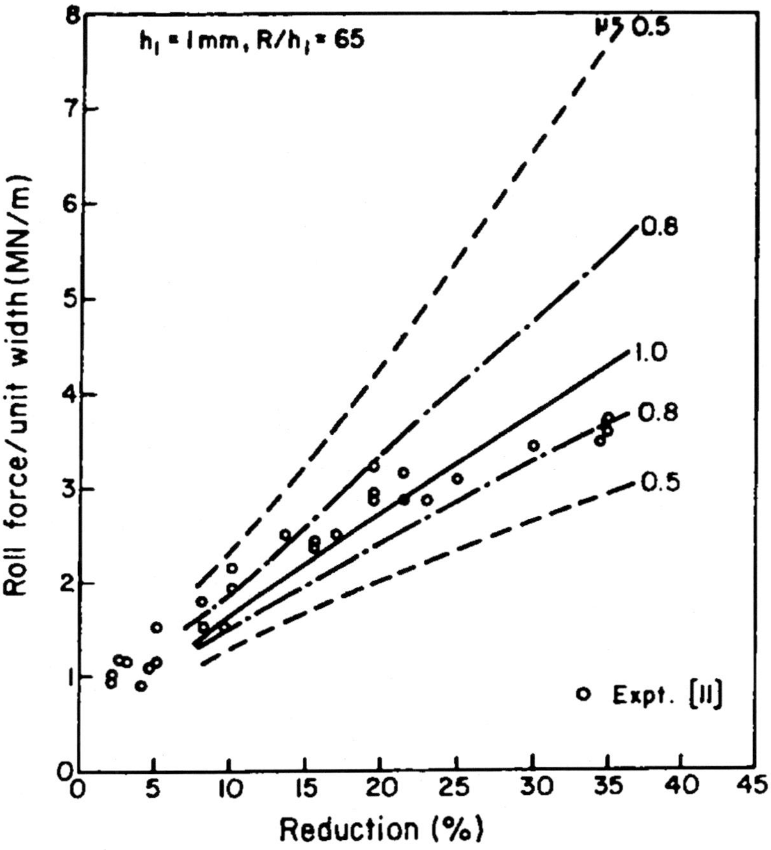

Figure 10. Variation of roll force with reduction at different membership grades. With permission from Dixit and Dixit [45]. Copyright [1996] Elsevier.

The method described above has been extended to the prediction of lower and upper estimates of the roll force and the roll torque based on the model described in [48]. The maximum approximation error for computing the coefficients for lower and upper prediction are termed as λ_1 and λ_2 respectively. The coefficients for lower and upper estimates are obtained by solving the following LP models:

$$\begin{aligned}
&\text{Minimize } \lambda_1,\\
&\text{subject to,}\\
&o_p - \hat{y}_p \le \lambda_1, \qquad p = 1, 2, \ldots\ldots.m,\\
&o_p - \hat{y}_p \ge 0, \qquad p = 1, 2, \ldots\ldots.m,\\
&\qquad\qquad \lambda_1 \ge 0,\\
&a_i^j \ge 0, \qquad i = 1, 2, \ldots\ldots.N_R, \quad j = 1, 2 \ldots..k.
\end{aligned} \tag{27}$$

and

$$
\begin{aligned}
&\text{Minimize } \lambda_2, \\
&\text{subject to,} \\
&-o_p + \hat{y}_p \le \lambda_2, \qquad p = 1, 2, \ldots\ldots m, \\
&o_p - \hat{y}_p \le 0, \qquad p = 1, 2, \ldots\ldots m, \\
&\qquad\qquad \lambda_2 \ge 0, \\
&a_i^j \ge 0, \qquad i = 1, 2, \ldots\ldots N_R, \quad j = 1, 2, \ldots\ldots k.
\end{aligned} \tag{28}
$$

Gudur and Dixit [47] have also proposed a method to remove the outliers from the supplied data for learning.

Figure 11 shows the fuzzification of 5 input variables used in [47]. The input variables are ratio of roll radius to inlet thickness (R/h_1), coefficient of friction (f), percentage reduction (r), and hardening parameters (b and n). The learning data from the fuzzy model is generated from a radial basis function neural network (NN) model.

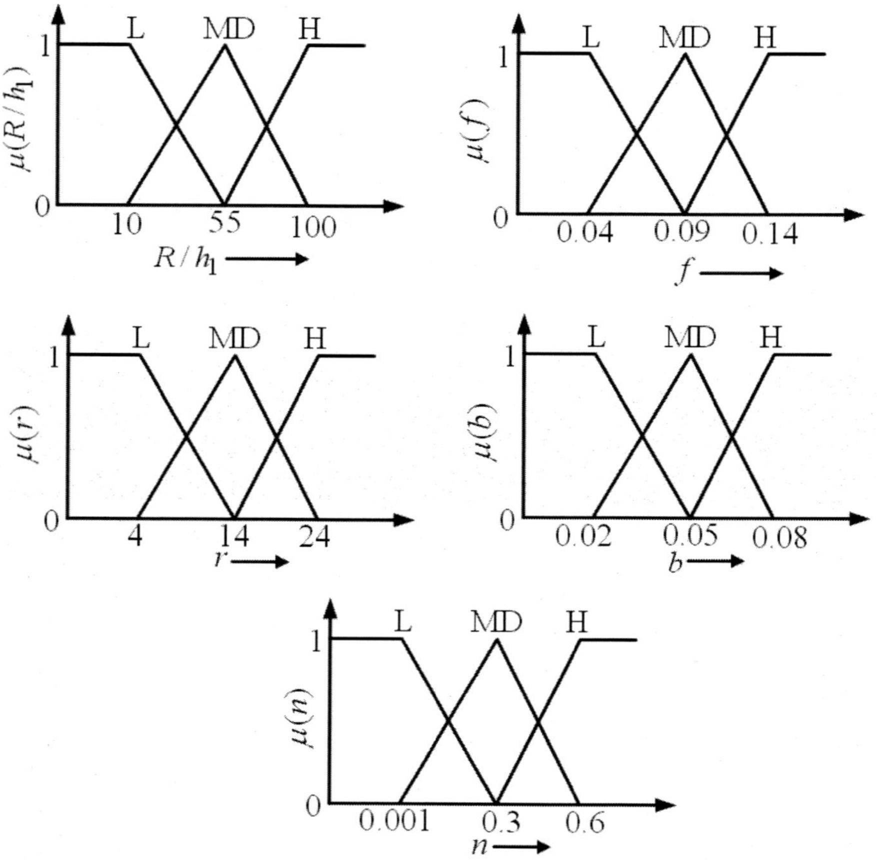

Figure 11. Fuzzification of input variables for a typical rolling process. With permission from Gudur and Dixit [2009]. Copyright (2009) Springer.

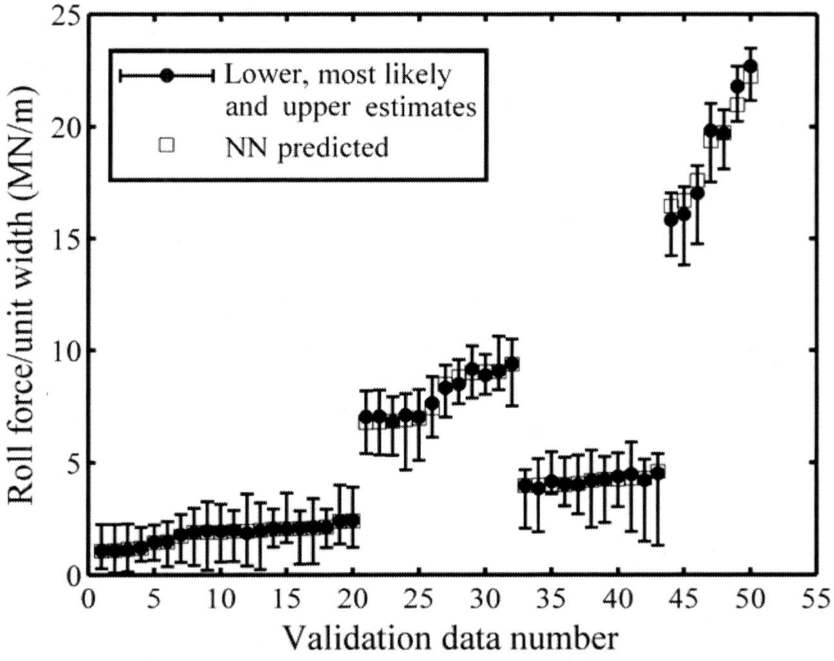

Figure 12. Lower, most likely and upper estimates of roll force. With permission from Gudur and Dixit [2009]. Copyright (2009) Springer.

The NN model in turn was trained by an FEM model. The lower, most likely and upper estimate of roll force is shown in Figure 12.

The fuzzy logic can be applied in the online control of gauge thickness of the sheet in the rolling process. It can also be used for controlling the flatness. Recently, Li and Janabi-Sharifi [49] has proposed a fuzzy logic based strategy for inter-stand tension control in a hot strip rolling mill. The control of tension is essential for maintaining the good quality of the product. The tension is controlled by regulating the speed of work rolls. Authors have used armature current of drive rolls as an indicator of the forward tension. Some other interesting papers describing the fuzzy set applications to other metal forming processes are [50–53].

Conclusion

This chapter has highlighted the applications of neural networks and fuzzy sets to machining and metal forming. Both these techniques have been applied in modeling, optimization and control of the processes. Each of these techniques can be used separately or together. For example Abburi and Dixit [31] used neural networks to learn from shop floor data; the neural network predicted data in turn were used to develop a fuzzy expert system. The hybrids of these methods have also been used, for example an adaptive-network-based fuzzy inference system (ANFIS) is a new soft computing tool that takes inspiration from fuzzy sets as well as neural networks. It uses the fuzzy rules of Sugeno type. The parameter of the membership functions and output functions are adaptively changed based on the input data. The training procedure is similar to neural networks and it uses a layered architecture similar to MLP network.

The neural network and fuzzy sets can also be used in conjunction with finite element model. Finite element model has established itself as a robust tool in modeling, however it takes a lot of computational time and does not provide a close form solution. Neural network, fuzzy sets and finite element method in combination can provide a fast, transparent and robust model.

For more successful applications of neural network and fuzzy sets in the machining and metal forming area, the following issues need the attention of the researchers:

- These methods require a lot of data for learning. Although a number of methods have been proposed to minimize the data required for learning, continuing effort is needed in this direction. The soft computing based methods should have the capability to learn from limited, imprecise and missing data.
- The robust data filtration algorithms should be developed to deal with the outliers in the data. The data may contain statistical variation. The soft computing based model should be able to predict the output in the form of a statistical variable. At present, barring a few exceptions, most of the models predict the mean value of the parameter.
- Fuzzy set theory is an excellent tool for incorporating the linguistic information in the model. However, construction of a suitable membership function for a variable is an art. There is a need to develop more systematic methods for it.
- Compatible hardware system like sensors, actuators *etc* should also be developed to take the maximum advantage of the soft computing methods.

REFERENCES

[1] Haykin, S. (2004), *Neural Networks— A Comprehensive Foundation,* Prentice-Hall of India, New Delhi.

[2] Klier, G.J. and Folger, T.A. (1993*), Fuzzy Sets, Uncertainty and Information*, Prentice-Hall of India, New Delhi.

[3] Bojadzieve, G. and Bojadzieve, M. (1995), Fuzzy Sets, Fuzzy Logic Applications, *Advances in Fuzzy Systems—Application and Theory*, Vol. 5, World Scientific, Singapore.

[4] Dixit, P.M. and Dixit, U.S. (2008), *Modeling of Metal Forming and Machining— By Finite Element and Soft Computing Methods*, Springer-Verlag, London.

[5] Kohli, A. and Dixit, U.S. (2005), A neural-network-based methodology for the prediction of surface roughness in a turning process, *International Journal of Advanced Manufacturing Technology,* Vol. 25, pp. 118–129.

[6] Sick, B. (2002), On-line and indirect tool wear monitoring in turning with artificial neural networks: a review of more than a decade of research, *Mechanical Systems and Signal Processing*, Vol. 16, pp. 487–546.

[7] Cus, F. and Zuperl, U. (2006), Approach to optimization of cutting conditions by using artificial neural networks, *Journal of Materials Processing Technology*, Vol. 173, pp. 281–290.

[8] Ko, T.J. and Cho, D.W. (1998), Adaptive optimization of face milling operations using neural networks, *ASME Journal of Manufacturing Science and Engineering*, Vol. 120, pp. 443–451.

[9] Liu, Y. and Wang, C. (1999), Neural network based adaptive control and optimization in the milling process, *International Journal of Advanced Manufacturing Technology*, Vol. 15, pp. 791–795.

[10] Azouzi, R. and Guillot, M. (1998), On-Line Optimization of the Turning Process Using an Inverse Process Neurocontroller, *ASME Journal of Manufacturing Science and Engineering,* Vol. 120, pp. 101–108.

[11] Pican, N., Alexandre, F. and Bresson, P. (1996), Artificial neural networks for the presetting of a steel temper mill, *IEEE Transactions: Expert*, Vol. 11, pp. 22–27.

[12] Larkiola, J., Myllykoski, P., Nylander, J. and Korhonen, A.S. (1996), Prediction of rolling force in cold rolling by using physical models and neural computing, *Journal of Materials Processing Technology*, Vol. 60, pp. 381–386.

[13] Cho, S., Cho Y. and Yoon, S. (1997), Reliable roll force prediction in cold mill using multiple neural networks, IEEE Transactions: *Neural Networks*, Vol. 8, pp. 874–882.

[14] Lee, D. and Lee, Y. (2002), Application of neural-network for improving accuracy of roll force model in hot-rolling mill, *Control Engineering Practice*, Vol. 10, pp. 473–478.

[15] Xie, H.B., Jiang, Z.Y., Liu, X.H. and Wang, G.D. and Tieu, A.K. (2006), Prediction of coiling temperature on run-out table of hot strip mill using data mining, *Journal of Materials Processing Technology*, Vol. 177, pp. 121–125.

[16] Geerdes, W.M., Alvarado, M. Á.T., Cabreara-Rios, M. And Cavazos, A. (2008), An application of physics-based and artificial neural networks-based hybrid temperature prediction schemes in a hot strip mill, *ASME Journal of Manufacturing Science and Engineering,* Vol. 130, 014501, pp. 1–5.

[17] Gorni, A.A. (1997), The application of neural networks in the modelling of plate rolling processes, *JOM-e,* Vol. 49, no. 4, April, electronic document.

[18] Gunasekera, J.S., Jia, Z., Malas, J.C. and Rabelo, L. (1998), Development of a neural network model for a cold rolling process, *Engineering Applications of Artificial Intelligence,* Vol. 11, pp. 597–603.

[19] Dixit, U.S. and Chandra, S. (2003), A neural network based methodology for the prediction of roll force and roll torque in fuzzy form for cold flat rolling process, *International Journal of Advanced Manufacturing Technology*, Vol. 22, pp. 883–889.

[20] Yang, Y.Y., Linkens, D.A., Talamantes-Silva, J. and Howard, I.C. (2003), Roll force and torque prediction using neural network and finite element modelling, *Transactions of the Iron and Steel Institute of Japan*, Vol. 43, pp. 1957–1966.

[21] Yang, Y.Y., Linkens, D.A., Talamantes-Silva, J. (2004), Roll load prediction— data collection, analysis and neural network modelling, *Journal of Materials Processing Technology,* Vol. 152, pp. 304–315.

[22] Osakada, A. and Yang G.B. (1991), Neural networks for process planning of cold forging, *CIRP Annals*, Vol. 40, pp. 243–246.

[23] Raj, K.H., Sharma, R.S., Srivastava, S. and Patvardhan, C. (2000), Modelling of manufacturing processes with ANNs for intelligent manufacturing, *International Journal of Machine Tools and Manufacture*, Vol. 40, pp. 851–868.

[24] Mori, T. and Li, S. (2005), Determination of design parameters in metal forging process by using FEM and neural network, *Proceedings of the 10th India-Japan Joint Seminar*, February 21–26, 2005, IIT Kanpur.

[25] Kazan, R., Firat, M. and Tiryaki, A.E. (2009), Prediction of springback in wipe bending process of sheet metal using neural network, *Materials and Design*, Vol. 30, pp. 418–423.

[26] Sbarbaro-Hofer, D., Neumerkel, D. and Hunt, K. (1993), Neural control of a steel rolling mill, *IEEE Transactions: Control Systems*, Vol. 13, pp. 69–75.

[27] Gudur, P.P. and Dixit, U.S. (2009), An application of fuzzy inference for studying the dependency of roll force and roll torque on process parameters in cold flat rolling, *International Journal of Advanced Manufacturing Technology*, Vol. 42, pp.41–52.

[28] Gudur, P.P. and Dixit, U.S. (2008), A neural network-assisted finite element analysis of cold flat rolling, *Engineering Applications of Artificial Intelligence*, Vol. 21, pp. 43–52.

[29] Panchal, J.H., Khanna, R. and Dixit, U.S. (2000), Optimization of turning process using a neuro-fuzzy controller, *Proceedings of Sixteenth National Convention of Mechanical Engineers and All India Seminar on Future Trends in Mechanical Engineering*, Research and Development, Roorkee, September 29-30, 2000.

[30] Liu, S.-T. (2006), Optimization of a machining economics model with fuzzy exponents and coefficients, *International Journal of Production Research*, Vol. 44, pp. 3083–3104.

[31] Abburi, N.R. and Dixit, U.S. (2006), A knowledge-based system for the prediction of surface roughness in turning process, *Robotics and CIM*, Vol. 22, pp. 363–372.

[32] Chen, J.C. and Black, J.T. (1997), A fuzzy-nets in-process (FNIP) systems for tool-breakage monitoring in end-milling operations, *International Journal of Machine Tools and Manufacture*, Vol. 37, pp. 783-800.

[33] Fang, X.D. and Jawahir, I.S. (1994), Predicting total machining performance in finish turning using integrated fuzzy-set models of the machinability parameters, *International Journal of Production Research*, Vol. 32, pp. 833–849.

[34] Fang, X.D., Fie, J. and Jawahir, I.S. (1996), A hybrid algorithm for predicting chip form/chip breakability in machining, *International Journal of Machine Tools and Manufacture*, Vol. 36, pp. 1093–1107.

[35] Hashmi K., El Baradie, M.A. and Ryan, M. (1999), Fuzzy-logic based intelligent selection of machining parameters, *Journal of Materials Processing Technology*, Vol. 94, pp. 94–111.

[36] Iqbal, A., He, N., Li, L. and Dar, N.U. (2007), A fuzzy expert system for optimizing parameters and predicting performance measures in hard-milling process, *Expert Systems with Applications* Vol. 32, pp. 1020–1027.

[37] Susanto, V. and Chen, J.C. (2003), Fuzzy logic based in-process tool wear monitoring system in face milling operations, *International Journal of Advanced Manufacturing Technology* Vol. 3, pp. 186–192.

[38] Ali, Y.M. and Zhang, L.C. (2004) A fuzzy model for predicting burns in surface grinding of steel, *Journal of Material Processing Technology*, Vol. 44, pp. 563–571.

[39] Liu, Q., Chen, X. and Gindy, N (2005) Fuzzy pattern recognition of AE signals for grinding burn, *International Journal of Machine Tools and Manufacture*, Vol. 45, pp. 811–818.

[40] Biglari, F.R. and Fang, X.D. (1995) Real-time fuzzy logic control for maximizing the tool life of small–diameter drills, *Fuzzy sets and systems*, Vol. 72, pp. 91–101.

[41] Chung B.M. and Tomizuka M. (2001) Fuzzy logic modeling and control for drilling of composite laminates, *10th IEEE International Conference on Fuzzy Systems*, Melbourne, pp. 509–512.

[42] Hashmi K., Graham I.D. and Mills B. (2000), Fuzzy logic based data selection for the drilling process, *Journal of Materials Processing Technology*, Vol.108, pp. 55–61.

[43] Dixit, U.S., Robi, P.S. and Sarma, D.K. (2002), A systematic procedure for the design of a cold rolling mill, *Journal of Materials Processing Technology*, Vol. 121, pp. 69–76.

[44] Dixit, U.S. and Dixit, P.M. (2000), Application of fuzzy set theory in scheduling of tandem rolling mills, *ASME Journal of Manufacturing Science and Engineering*, Vol. 122, pp. 494–500.

[45] Dixit, U.S. and Dixit, P.M. (1996), A finite element analysis of flat rolling and application of fuzzy set theory. *International Journal of Machine Tools and Manufacture,* Vol. 36, pp. 947–969.

[46] Shida, S. and Awazuhara, H. (1973), Rolling load and torque in cold rolling, *Journal of the Japan Society for Technology of Plasticity*, Vol. 14, pp. 267–278.

[47] Gudur, P.P. and Dixit, U.S. (2009), An application of fuzzy inference for studying the dependency of roll force and roll torque on process variables in cold flat rolling, *International Journal of Advanced Manufacturing Technology*, Vol. 42, pp. 41–52.

[48] Skrjanc, I., Blazic, S. and Agamennoni, O. (2005), Interval fuzzy model identification using l_∞-norm, *IEEE Transactions: Fuzzy Systems*, Vol. 13, pp. 561–568.

[49] Li, G. and Janabi-Sharifi, F. (2009) Fuzzy looperless tension control for hot strip rolling, *Fuzzy Sets and Systems*, Vol.160, pp. 521–536.

[50] Ong, S.K, De Vin, L.J., Nee, A.Y.C. and Kals, H.J.J. (1997), Fuzzy set theory applied to bend sequencing for sheet metal bending, *Journal of Materials Processing Technology,* Vol. 69, pp. 29–36.

[51] Osakada, K., Yang, G. and Mori, K. (1993), Determination of Optimum Forming Path in Three-Roll Bending by Combination of Fuzzy Reasoning and Finite Element Simulation, *CIRP Annals— Manufacturing Technology*, Vol. 42, pp. 291–294.

[52] Hsiang, S.H. and Lin Y.W. (2008), Application of fuzzy theory to predict deformation behaviors of magnesium alloy sheets under hot extrusion, *Journal of Materials Processing Technology*, Vol. 201, pp.138–144.

[53] Shiraishi, M., Nikawa, M., Kubota, T. and Goto (2007), Y., Prediction of curvatures of parts extruded through rotary die using fuzzy inference, *Journal of Materials Processing Technology,* Vol.187-188, pp. 702–705.

In: Artificial Intelligence in Manufacturing Research
Editor: J. Paulo Davim
ISBN 978-1-62618-557-9

Chapter 2

MULTI-OBJECTIVE OPTIMIZATION OF MULTI-PASS MILLING PROCESS PARAMETERS USING ARTIFICIAL BEE COLONY ALGORITHM

R. Venkata Rao[1]* and P. J. Pawar[2]
[1]Department of Mechanical Engineering,
S.V. National Institute of Technology,
Ichchanath, Surat, Gujarat, India
[2]Department of Production Engineering,
K.K. Wagh Institute of Engineering Education and Research
Nashik, Maharashtra, India

ABSTRACT

The effective optimization of machining process parameters affects dramatically the cost and production time of machined components as well as the quality of the final products. This chapter presents the details of multi-objective optimization of a multi-pass milling operation. The two objectives considered are maximization of production rate and minimization of production cost subjected to various constraints of arbor strength, arbor deflection, cutting power, and surface roughness.

Various cutting strategies are considered to determine the optimal process parameters like the number of passes, depth of cut for each pass, cutting speed, and feed per tooth. The optimization is carried out using a recently developed non-traditional optimization algorithm namely, artificial bee colony (ABC) algorithm. An example is presented and solved to illustrate the effectiveness of the algorithm. The results show that the artificial bee colony algorithm can be effectively used in the multi-objective optimization of multi-pass milling process.

* Corresponding author, E-mail: ravipudirao@gmail.com

INTRODUCTION

In today's manufacturing environment many large industries have attempted to introduce the flexible manufacturing system (FMS) as a strategy to adapt to the ever-changing competitive market requirement. The flexible manufacturing system involves highly automated and computer controlled machines. Due to high capital and machining costs, there is an economic need to operate these machines as efficiently as possible in order to obtain the required pay back. The success of any machining operation depends on the selection of machining process parameters. Proper selection of process parameters play a significant role to ensure quality of product, to reduce the machining cost, to increase productivity in computer controlled machining processes and to assist in computer aided process planning. A human process planner selects machining process parameters using his own experience or from the handbooks. But these parameters do not give optimal result. Various optimization strategies and algorithms ranging from elementary numerical search methods to more systematic approaches employing non-traditional techniques for optimization of process parameters in case of single pass milling operation had been reported in the literature. However, as multi-pass operations are often preferred to single pass operations for economic reasons, recent efforts have been directed towards determination of optimal machining conditions for multi-pass operations.

Traditionally, mathematical programming techniques like linear programming, method of feasible direction, dynamic programming and geometric programming had been used to solve optimization problems in milling. However, these traditional methods of optimization do not fare well over a broad spectrum of problem domains. Moreover, traditional techniques may not be robust. Numerous constraints and multiple passes make machining optimization problems complicated and hence these techniques are not ideal for solving such problems as they tend to obtain a local optimal solution.

Considering the drawbacks of traditional optimization techniques, attempts are being made to optimize the machining problem using evolutionary optimization techniques. Evolutionary computation consists of a variety of methods including optimization paradigms that are based on evolution mechanisms such as biological genetics and natural selection. These methods use the fitness information instead of the functional derivatives making them more robust and effective. These methods thus avoid the problem of getting trapped in local optima and enable to obtain a global (or nearly global) optimum solution. Efforts are continuing to use more recent optimization algorithms, which are more powerful, robust and able to provide accurate solution. Artificial bee colony (ABC) algorithm developed by Karaboga [1] and Karaboga and Basturk [2, 3] is one of the recent algorithms and no effort has been yet made for optimization of process parameters of any of the machining processes by using this algorithm. This chapter presents an application of artificial bee colony algorithm (ABC) for multi-objective optimization of process parameters of multi-pass milling operation.

Milling is the machining process in which the metal is removed by a rotating multiple tooth cutter. Milling operation can be performed in a single pass or in multi-passes. Multi-pass operations are often preferred to single pass operations for economic reasons and are generally used to machine stocks that cannot be removed in a single pass. In case of face milling operation a significant amount of stock material is to be removed and hence work piece is required to be machined using multi-pass operations. Determination of the optimal cutting parameters like number of passes, depth of cut for each pass, speed, and feed is considered crucial of multi-pass milling operation. Various investigators had used optimization techniques, both traditional and non-traditional, for optimization of multi-pass milling operation in the past.

Shin and Joo [4] used dynamic programming optimization method for milling process parameter optimization. Dynamic programming can solve both continuous and discrete variables and yield a global optimal solution. However, if the optimization problem involves a large amount of independent parameters with a wide range of values such as the cutting parameters in milling operation, the use of dynamic programming is limited. Wang [5] used a neural network based approach to optimize milling process parameters. However, optimization by using neural networks may often ends in local minima or fails to converge on a desired result.

Tolouei-Rad and Bidhendi [6] used the method of feasible direction and considered maximization of profit rate as an objective function in milling operation. The feasible solution denotes the local minimum of the problem. However, this local minimum need not be the global one unless the problem is convex programming problem. Optimization model developed in their work was non-convex.

Sonmez et al. [7] studied multi-pass milling operation based on the maximum production rate criterion and used an algorithm adopted from the study of Agapiou [8] which was proposed for the multi-pass turning operations. The authors had developed a strategy to determine the optimum cutting parameters for multi-pass milling based on the maximum production rate criterion. The optimum number of passes was determined via dynamic programming, and the optimal values of the cutting conditions were found based on the objective function developed for the typified criterion by using geometric programming. Although the results showed significant improvement over handbook recommendations, the optimization techniques used in their work either tend to result in local minima or take a long time to converge on a reasonable result.

Shunmugam et al. [9] used genetic algorithm (GA) for milling process parameter optimization with total production cost as the objective function. The authors had divided the cost into two separate minimization problems of roughing and finishing operations. GA was used for finding the optimum number of rough passes and allocation of the total stock in each of the rough and the finish passes to achieve the minimum total production cost. However, although GA has advantages over the traditional techniques, it has the drawbacks of lack of ability to perform good local search and premature convergence.

During the past decade, different optimization methods had been integrated to improve performance of algorithms and to reach the global optimum results. In order to optimize the machining parameters, the evolutionary methods had been modified or hybridized by using other optimization techniques. Liu and Wang [10] modified the genetic algorithm by defining and changing the operating domain and used for optimization of milling parameters. The results and the convergence speed of their approach were better than that of genetic algorithm. Wang et al. [11] proposed a new hybrid approach, named genetic simulated annealing (GSA), based on genetic algorithm and simulated annealing to find optimal machining parameters in milling operations. In their approach, genetic algorithm and simulated annealing were combined. They pointed out that the results obtained were found to be better than those of genetic algorithm and geometric programming.

Baskar et al. [12] considered a specific case in milling operation and solved the same by using three different non-traditional optimization techniques comprising a global search algorithm (genetic algorithm), local search algorithm (local hill climbing) and memetic algorithm. Significant improvement was obtained with the above techniques in comparison with the results given by handbook and the method of feasible direction. This was because of the limitations of enumerative search followed by the direct search method.

Onwubolu [13] proposed a new optimization technique based on Tribes for determination of the cutting parameters in multi-pass milling operations such as plain milling and face milling by simultaneously considering multi-pass rough machining and finish machining. The optimum milling parameters were determined by the maximum production rate criterion subjected to several practical technological constraints. The results obtained confirmed the well known fact that multi-pass approach is preferred to single-pass approach if the total depth of cut to be removed in milling exceeds the maximum allowable depth of cut. Although the results obtained in his work using tribes showed significant improvement over other traditional and non-traditional algorithms, but the results are not valid as some of the constraints in the solution obtained are violated.

Yildiz [14] developed a new hybrid optimization approach by hybridizing the immune algorithm with hill climbing local search algorithm to maximize the total profit rate in milling operations. The results showed that the hybrid approach was more effective to optimize the cutting parameters for milling operations than genetic algorithm, the feasible direction method, and the handbook recommendations. Zarei et al. [15] presented a harmony search (HS) algorithm to determine the optimum cutting parameters for multi-pass face-milling. Total production cost was considered as the objective function.

It is also revealed from the literature that few efforts has been made for multi-objective optimization of some machining processes like drilling [16], grinding [17, 18] and turning [19]. However, no effort has been yet made for multi-objective optimization of multi-pass milling process.

The literature related to optimization of milling process is thus mainly concerned with single objective optimization only (considering either production cost or production time), with single pass or multi-pass operation. This chapter therefore provides an approach to develop a methodology for multi-objective optimization of multi-pass milling process with optimization using artificial bee colony algorithm. The two objectives considered are (1) maximization of production rate (i.e. minimization of production time) and (2) minimization of production cost. Although a low production time would mean a low production cost, it should be realized that machining parameters giving minimum production time would not be identical to those giving minimum production cost. Hence it is necessary to determine the compromise best solution which would ultimately lead to highest benefit in terms of both cost and productivity. Various practical constraints considered are, arbor strength, arbor deflection, cutting power, and surface roughness. Feed per tooth, cutting speed and depth of cut are considered as process parameters. The upper and lower bounds of the process parameters are also included in the study.

The next section presents a multi-objective optimization model of multi-pass milling process.

MULTI-OBJECTIVE OPTIMIZATION MODEL OF MULTI-PASS MILLING PROCESS

The objective functions and the constraints are formulated as discussed below.

Objective Functions

Two objectives considered are:

(1) Maximization of production rate (i.e. minimization of production time)
(2) Minimization of production cost

Maximization of Production Rate

Production rate is maximized by minimizing the production time. Equation (1) gives the production time for multi-pass milling operation [7].

$$T_{\mathrm{Pr}} = T_P + T_L + N_p T_a + \sum_{i=1}^{N_p} \frac{\pi DL}{f_{zi} z 1000 V_i} + \frac{T_d \pi L V_i^{\left(\frac{1}{m}-1\right)} a_i^{e_v/m} f_{zi}^{\left(\frac{u_v}{m}-1\right)} a_r^{r_v/m} z^{\left(\frac{n_v}{m}-1\right)} \lambda_s^{q_v/m}}{1000 C_v^{1/m} D^{\left(\frac{b_v}{m}-1\right)} \times \left(B_m B_h B_p B_t\right)^{1/m}} \quad (1)$$

where, T_P = machine preparation time; T_L = Loading and unloading time; N_p = Number of passes; T_a = Process adjusting and quick return time; D = cutter diameter; L= length of cut; T_d = Tool changing time; f_z = feed per tooth; z = Number of teeth; V = cutting speed; a = depth of cut, a_r = width of the cut, B_m, B_k, B_p, B_t = correction coefficients, m, e_v, u_v, r_v, n_v, q_v, b_v, = exponents, C_v = process constant, λ_s = cutting inclination angle.

Also, let,

$$T_{\mathrm{Pr}} = T_1 + T_2 \quad (2)$$

where,

$$T_1(\min) = \frac{T_s}{N_b} + T_L + N_p T_a \quad (3)$$

$$T_2(\min) = \sum_{i=1}^{N_p} \frac{\pi DL}{f_{zi} z 1000 V_i} + \frac{T_d \pi L V_i^{\left(\frac{1}{m}-1\right)} a_i^{e_v/m} f_{zi}^{\left(\frac{u_v}{m}-1\right)} a_r^{r_v/m} z^{\left(\frac{n_v}{m}-1\right)} \lambda_s^{q_v/m}}{1000 C_v^{1/m} D^{\left(\frac{b_v}{m}-1\right)} \times \left(B_m B_h B_p B_t\right)^{1/m}} \quad (4)$$

Minimization of Production Cost

Equation (5) gives the production cost in multi-pass milling operation [9].

$$U_t = U_f + \sum_{i=1}^{N_p} U_{ri} + A_4 \tag{5}$$

where, A_4 = Cost of tool preparation, = $k_o T_p$; k_o = Overhead cost,

Cost per pass is given by (U)

$$U(\$) = k_o t_m + (k_t z)\frac{t_m}{T_R} + k_o t_m\left(\frac{zT_d}{T_R}\right) + k_o(Lh_1 + h_2) \tag{6}$$

$$\text{Machining time} = t_m = \frac{\pi DL}{1000 z f_z V} \tag{7}$$

k_t = cost of cutting edge;

$$T_R = \text{tool replacement life} = \frac{C_v^{1/m} D^{\left(\frac{b_v}{m}\right)} \times \left(B_m B_h B_f B_t\right)^{1/m}}{V^{\left(\frac{1}{m}\right)} a^{e_v/m} f_z^{\left(u_v/m\right)} a_r^{r_v/m} z^{\left(\frac{n_v}{m}\right)} \lambda_s^{q_v/m}} \tag{8}$$

h_1 = tool returns time; h_2 = Tool advance/return time;

Constraints

Following four constraints are considered in this optimization model:

Arbor Strength

The arbor is subjected to torsion from the action of resistance to cutting. Therefore, the selected values of process parameters should ensure that the arbor is safe from strength point of view as specified by equation (9).

$$F_s - F_c \geq 0 \tag{9}$$

where, F_c = mean peripheral cutting force $= C_{zp} a_r z D^{b_z} a^{e_z} f_z^{u_z}$ (10)

C_{zp}=, process constants, b_z, e_z, and u_z are exponents.

Permissible force for arbor strength (Kg) = F_s

$$F_s = \frac{0.1k_b d_a^3}{0.08L_a + 0.65\sqrt{\left((0.25L_a)^2 + (0.5\alpha D)^2\right)}} \tag{11}$$

where, k_b = permissible bending strength of arbor; d_a = arbor diameter; L_a = arbor length between supports; $\alpha = k_b / (1.3\ k_t)$; k_t = permissible torsional strength of arbor.

Arbor Deflection

The selected values of process parameters should be checked for arbor deflection as given by equation (12).

$$F_d - F_c \geq 0 \tag{12}$$

where, permissible force for arbor deflection (Kg) = $F_d = \frac{4Eed_a^4}{L_a^3}$ (13)

where, E = modulus of elasticity of arbor material; e = permissible value of arbor deflection.

For rough milling operation, e=0.2 mm and for finish milling operation, e=0.05 mm.

Power

Power required for the cutting operation should not exceed the effective power transmitted to cutting point by the machine tool. This is ensured by equation (14).

$$P_c - \frac{F_c V}{6120} \geq 0 \tag{14}$$

where, P_c = Cutting power (KW) = $P_m * \eta$

P_m = nominal motor power, η=overall efficiency.

Surface Roughness Constraint

For roughing operation the surface roughness value should not exceed the specified limit of 25 μm as given by equation (15).

$$25 \times 10^{-3} - \frac{0.0321 f_z^2}{r_e} \geq 0 \tag{15}$$

where, r_e = nose radius = 0.999 mm

For finishing operation the surface roughness value should not exceed the specified limit of 2.5 μm as given by equation (16).

$$2.5 \times 10^{-3} - \frac{0.0321 f_z^2}{r_e} \geq 0 \tag{16}$$

Process Variables

The three process variables considered are depth of cut (a), feed per tooth (f_z) and cutting speed (V). The bounds for these variables are given by equations below.

$$1 \leq a \leq 4 \text{ (mm)} \quad (17)$$
$$0.1 \leq f_z \leq 0.6 \text{ (mm/tooth)} \quad (18)$$
$$50 \leq V \leq 300 \text{ (m/min)} \quad (19)$$

The next section describes the artificial bee colony algorithm as an optimization technique.

ARTIFICIAL BEE COLONY ALGORITHM

A branch of nature inspired algorithms, called swarm intelligence, is focused on insect behavior in order to develop some meta-heuristics which can mimic insect's problem solution abilities. Interaction between insects contributes to the collective intelligence of the social insect colonies.

These communication systems between insects have been adapted to scientific problems for optimization. The foraging behavior, learning, memorizing and information sharing characteristics of honeybees have recently been one of the most interesting research areas in swarm intelligence. Artificial bee colony (ABC) algorithm is developed to model the intelligent behaviors of honeybee swarms .The honeybee swarms consists of two essential components (i.e. food sources and foragers) and defines two leading modes of the behavior (i.e. recruitment to a nectar source and abandonment of a source).

Food Sources

The value of a food source depends on different parameters such as its proximity to the nest, richness of energy and ease of extracting this energy. For the simplicity, the ''profitability'' of a food source can be represented with a single quantity.

Foragers

Foragers can be unemployed, employed or experienced.

Unemployed Foragers

If it is assumed that a bee has no knowledge about the food sources in the search field, bee initializes its search as an unemployed forager. There are two possibilities for an unemployed forager:

- Scout Bee: If the bee starts searching spontaneously without any knowledge, it will be a scout bee. The percentage of scout bees varies from 5% to 30% according to the information into the nest. The mean number of scouts averaged over conditions is about 10%.
- Recruit: If the unemployed forager attends to a waggle dance done by some other bee, the bee will start searching by using the knowledge from waggle dance.

Employed Foragers

When the recruit bee finds and exploits the food source, it becomes an employed forager and memorizes the location of the food source. After the employed foraging bee loads a portion of nectar from the food source, it returns to the hive and unloads the nectar to the food area in the hive. There are three possible options related to residual amount of nectar for the foraging bee. If the nectar amount decreased to a low level or exhausted, foraging bee abandons the food source and become an unemployed bee. If there are still sufficient amount of nectar in the food source, it can continue to forage without sharing the food source information with the nest mates or it can go to the dance area to perform waggle dance for informing the nest mates about the same food source. The probability values for these options highly related to the quality of the food source.

Experienced Foragers

These types of foragers use their historical memories for the location and quality of food sources. This type of forages can be an inspector, which controls the recent status of food source already discovered. It can also be a reactivated forager by using the information from waggle dance. It tries to explore the same food source discovered by self if there are some other bees confirm the quality of same food source. It can also be scout bee to search new patches if the whole food source is exhausted. It can also be a recruit bee, which is searching a new food source declared in dancing area by another employed bee.

Communication among bees related to the quality of food sources occurs in the dancing area. The related dance is called waggle dance. Since information about all the current rich sources is available to an onlooker on the dance floor, she probably could watch numerous dances and choose to employ herself at the most profitable source. There is a greater probability of onlookers choosing more profitable sources since more information is circulating about the more profit able sources. Employed foragers share their information with a probability, which is proportional to the profitability of the food source, and the sharing of this information through waggle dancing is longer in duration. Hence, the recruitment is proportional to profitability of a food source. The detailed steps of artificial bee colony algorithm are explained in section 5.

The nest section provides an example of multi-objective multi-pass milling process to demonstrate and validate the application of the artificial bee colony algorithm.

EXAMPLE

Now, the following experimental data are considered to demonstrate and validate the application of artificial bee colony algorithm for the optimization of process parameters of the multi-pass milling operation. The data considered for the example are as given below:

Motor power (Pm) = 5.5 kW, efficiency, $\eta = 0.7$
Arbor diameter, $d_a = 24$ mm, arbor length between supports, $L_a = 210$ mm
Permissible bending stress of arbor, k_b: 100 MPa
Permissible torsional stress of arbor, k_t: 85 MPa
Modulus of elasticity of arbor material, $E = 200$ GPa,
Tool material: HSS, tool diameter, $D = 120$ mm, number of teeth, $z = 8$,
Material: structural carbon steel
Tensile strength: 750 MPa, Brinell hardness number = 150
Length of cut, $L = 200$ mm, width of cut, $a_r = 50$ mm, depth of cut, $a = 5$ mm
Machine preparation time, $T_p = 0.75$ min.
Loading and unloading time of one work piece, $T_L = 1.5$ min.
Tool change time, $T_d = 5$ min.
Process adjusting and quick return time, $T_a = 0.1$ (min/part)
Cutting inclination $\lambda_s = 30°$
Overhead cost, $k_o = 0.5$ ($/min), Cost of cutting edge, $k_t = 2.5$ ($/cut edge)
Tool returns time, $h_1 = 7\times10{-}4$ min/mm, Tool advance time, $h_2 = 0.3$ min
Constants: B_m=1, B_k=1, B_p=0.8, B_t=0.8, m=0.33, e_v=0.3, u_v=0.4, r_v=0.1, n_v=0.1, q_v=0, C_v=35.4, b_v=0.45, $C_{zp} = 68.2$, $b_z = -0.86$, $e_z = 0.86$, and $u_z = 0.72$.

Now various steps of ABC algorithms are applied as described below:

Step 1: Parameter Selection

As discussed in the description of ABC algorithm, food source represents a possible solution to the problem of multi-objective optimization with minimization of production time and minimization of cost as considered in this chapter. Number of initial solutions (i.e. the number of food sources) considered for the given problem are five. The value of each food source depends on the fitness value of the objective function given by equations (1) and (5).

For every food source there is only one employed bee (employed forager). In other words, the number of employed bees is equal to number of food sources. Thus, the number of employed bees is considered to be *five*. The unemployed forager can be scout or an onlooker bee. The number of onlooker bees must be greater than the number of employed bees. The effect of number of onlooker bees on the convergence of solution is shown in Figure 1. It is observed that as the number of onlooker bees (n_0) and hence the population size increases, the algorithm perform better in terms of convergence rate. However, after a sufficient value of number of onlooker bees, any increment in the value does not improve the performance of the algorithm.

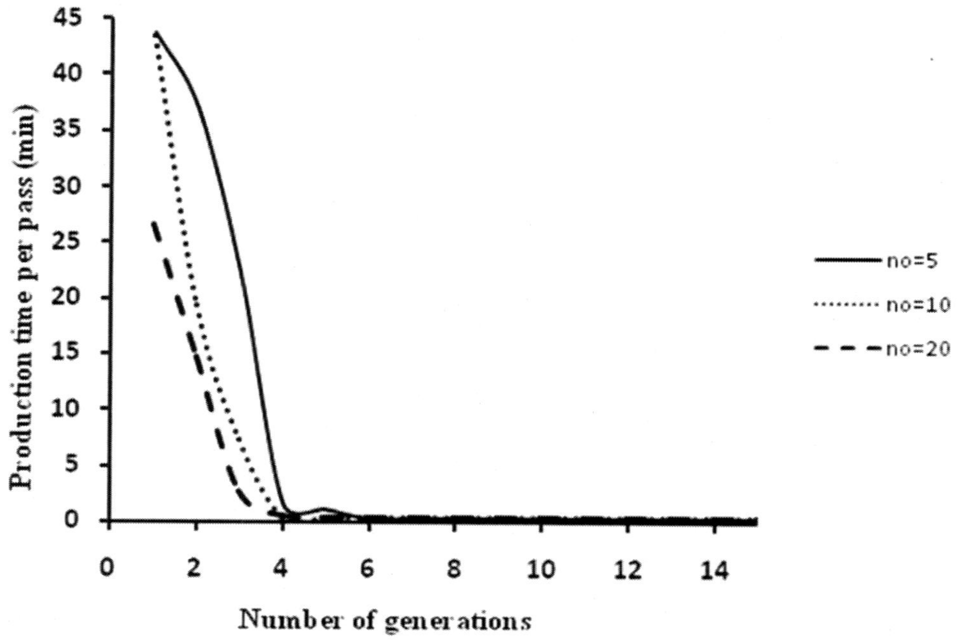

Figure 1. Effect of number of onlooker bees (no) on convergence rate for rough milling.

For the problem considered in this chapter, number of onlooker bees is considered to be *eleven*, which can provide an acceptable convergence speed for search. The colony size is the sum of number of employed bees and number of onlooker bees. Hence the colony size is *sixteen*. Number of scout bees is usually 5-30% of the colony size. The number of scout bee is taken as 5% of the colony size i.e. *one*. The parameters of optimization thus selected are summarized as below:

- Number of employed bees=5
- Number of onlookers bees= 11
- Number of scout bees=1
- Maximum number of iterations= 150

Step 2: Calculate the Nectar Amount of Each Food Source

The employed bees are moved to the food sources and the nectar amount of these food sources is evaluated based on their fitness value as defined by the objective function given by equation (1) and (5) subjected to constraints given by equations (9),(12), (14), (15) and (16).

Step 3: Determine the Probabilities by Using the Nectar Amount

If the nectar amount of a food source "θ_i" is F_i, then the probability (P_i) of choosing this food source by an onlooker bee is expressed as:

$$P_i = \frac{\sum_{k=1}^{S}(1/f_k)^{-1}}{f_i} \tag{20}$$

where "S" is the number of food sources.

Step 4: Calculate the Number of Onlooker Bees, Which Will Be Sent to Food Sources

Based on the probabilities calculated in step 3, the number (*N*) of onlookers bees sent to food source "θ_i" is calculated as:

$$N = P_i * m. \tag{21}$$

where, 'm' is the total number of onlooker bees.

Step 5: Calculate the Fitness Value of Each Onlooker Bee

After watching the dances of employed bees, an onlooker bee goes to the region of food source "θ_i" by the probability given by equation (20). The position of the selected neighbor food source is calculated as the shown in equation (22).

$$\theta_i\,(c+1) = \theta_i\,(c) \pm \phi_i\,(c) \tag{22}$$

where, "c" is number of generation. ϕ_i (c) is a randomly produced step to find a food source with a more nectar around 'θ_i'. ϕ_i (c) is calculated by taking the difference of the same parts of θ_i(c) and θ_k(c) ("k" is a randomly produced index) food positions. If the nectar amount F_i(c+1) at θ_i(c+1) is higher than at θ_i(c), then the bees go to the hive and share information with others and the position θ_i(c) of the food source is changed to θ_i(c+1) otherwise θ_i(c) is kept as it is. If the position 'θ_i' of the food source "i" cannot be improved through the predetermined number of trials, then that food source 'θ_i' is abandoned by its employed bee and then the bee becomes a scout. The scout starts searching new food source, and after finding the new source, the new position is accepted as 'θ_i'.

Step 6: Evaluate the Best Solution

Position of the best onlooker bee is identified for each food source. The global best of the honeybee swarm in each generation is obtained and it may replace the global best at previous generation if it has better fitness value.

Step 7: Update the Scout Bee

The worst employed bees, as many as the number of scout bees in the population, are respectively compared with the scout solutions. If the scout solution is better than employed solution, employed solution is replaced with scout solution. Else employed solution is transferred to the next generation without any change.

Using ABC algorithm, various feasible cutting strategies are adopted to determine the optimum number of passes required (i.e. number of roughing and finishing passes) and depth of cut for each pass.

The results of optimization are described below:

1. Determination of optimum process parameters of finish milling considering only production rate as the objective.

Table 1 shows the results of optimization of finish milling operation considering only production rate as the objective. It may be observed that all the constraints are satisfied. For finish milling, values of depth of cut higher than 1.5 are not considered as for these values the deflection constraint is violated for any combination of feed and speed in specified range.

Table 1. The results of optimization of finish milling operation considering only production rate as the objective

a	fz	V	T2	SC	DC	PC	RC
1	0.213	92.727	0.711	64.0751	0.0751	1.6390	0.001044
1.5	0.131	100.189	1.071	64.2630	0.2630	1.4448	0.001949

a: depth of cut (mm); fz: feed per tooth (mm/tooth); V: cutting speed (m/min);

SC, DC, PC and RC are the values of left hand side of equations (9), (12), (14) and (16) related to strength, deflection, power and roughness constraints respectively.

2. Determination of optimum process parameters of finish milling considering only production cost as the objective

Table 2 shows the results of optimization of finish milling operation considering only production cost as the objective.

Table 2. The results of optimization of finish milling operation considering only cost as the objective

a	fz	V	U	SC	DC	PC	RC
1	0.212	50	1.199	64.5688	0.5687	2.6618	0.00106
1.5	0.130	50	1.639	65.0649	1.0649	2.6658	0.00196

3. Multi-objective optimization of finish milling

The combined objective function of finish milling (Z_F) is formulated as given by equation (23).

$$Z_F = w_1 \times \frac{T_2}{d_1} + w_2 \times \frac{U}{d_2} \tag{23}$$

where, w_1 and w_2 are the weights assigned to the objectives of production rate and production cost, respectively. The values of weights can be calculated by using the analytic hierarchy process [20]. However, in the present example, equal weights are assumed. d_1 and d_2 are the minimum values of production time and production cost respectively for a given depth of cut, obtained in step 1 and step 2 above. Table 3 shows the results of multi-objective optimization for finish milling operation for a=1 mm {d_1=0.711; d_2=1.199} and for a=1.5 mm{d_1=1.071; d_2=1.639}.

Table 3. The results of multi-objective optimization for finish milling operation

a	fz	V	T2	U	ZF	SC	DC	PC	RC
1	0.211	57.08	0.869	1.311	1.156	65.063	1.063	2.498	0.0010
1.5	0.131	59.58	1.332	1.821	1.174	64.263	0.263	2.431	0.0019

4. Determination of optimum process parameters of rough milling considering only production rate as the objective

Table 4 shows the results of optimization of rough milling operation considering only production rate as the objective. It may be observed that all the constraints are satisfied. For rough milling, values of depth of cut higher than 2.5 are not considered as for these values the strength constraint is violated for any combination of feed and speed in specified range.

Table 4. The results of optimization of rough milling operation considering only production rate as the objective

a	fz	V	T2	SC	DC	PC	*RC*
1	0.353	50.00	0.608	0.0607	374.061	2.1348	0.0210
1.5	0.217	81.28	0.794	0.0552	374.055	1.0617	0.0234
2	0.154	85.91	1.064	0.3008	374.301	0.9063	0.0242
2.5	0.118	89.34	1.335	0.2609	374.261	0.7881	0.0245

a: depth of cut (mm); fz: feed per tooth (mm/tooth); V: cutting speed (m/min);
SC, DC, PC and RC are the values of left hand side of equations (9), (12), (14) and (15) related to strength, deflection, power and roughness constraints respectively.

5. Determination of optimum process parameters of rough milling considering only production cost as the objective

Table 5 shows the results of optimization of rough milling operation considering only production cost as the objective.

Table 5. The results of optimization of rough milling operation considering only cost as the objective

a	fz	V	U	SC	DC	PC	RC
1	0.352	50.00	1.083	0.489	374.489	2.138	0.021
1.5	0.217	50.00	1.431	0.402	374.403	2.137	0.023
2	0.154	50.00	1.771	0.300	374.301	2.136	0.024
2.5	0.117	50.00	2.109	1.542	375.542	2.146	0.024

6. Multi-objective optimization of rough milling

Now, the combined objective function for rough milling is formulated as given by equation (24) which is similar to equation (23).

$$Z_R = w_1 \times \frac{T_2}{d_1} + w_2 \times \frac{U}{d_2} \tag{24}$$

In case of rough milling, Table 6 shows the results of multi-objective optimization for rough milling operation for a=1mm $\{d_1=0.608;\ d_2=1.083\}$, for a=1.5mm$\{d_1=0.794;\ d_2=1.431\}$; for a=2mm: $\{d_1=1.064;\ d_2=1.771\}$ and for a=2.5mm: $\{d_1=1.335;\ d_2=2.109\}$.

Table 6. The results of multi-objective optimization for rough milling operation

a	fz	V	T2	U	ZR	SC	DC	PC	R
1	0.352	50.00	0.610	1.084	1.000	0.489	374.489	2.1380	0.0210
1.5	0.217	50.00	0.965	1.432	1.106	0.402	374.402	2.1376	0.0234
2	0.154	52.91	1.288	1.852	1.127	0.300	374.300	2.0370	0.0242
2.5	0.117	55.41	1.620	2.284	1.145	1.542	375.542	1.9620	0.0245

7. Formulation of equivalent objective function (Z_{eq}) for finish milling and rough milling operations

The equivalent objective function Z_{eq} is written as:

$$Z_{eq} = (T_2)_{eq} + (U)_{eq} \tag{25}$$

where $(T_2)_{eq} = T_2/\,(T_{2min})_{a=1};\ (U)_{eq} = U/(U_{min})_{a=1}$

The values of equivalent objective function for various depths of cut in finish milling operation are given in Table 7 and that for rough milling operation are given in Table 8.

Table 7. Equivalent objective function for finish milling

a	fz	V	(T2)eq	(U)eq	(Zeq)finish
1	0.211	57.08	1.222	1.092	2.314
1.5	0.131	59.58	1.873	1.517	3.390

Table 8. Equivalent objective function for rough milling

a	fz	V	(T2)eq	(U)eq	(Zeq)finish
1	0.352	50.00	1.000	1.000	2.000
1.5	0.217	50.00	1.587	1.321	2.908
2	0.154	52.91	2.118	1.708	3.826
2.5	0.117	55.41	2.664	2.107	4.771

8. Determination of total equivalent objective function

The total equivalent objective function $(Z_{eq})_{total}$ is then obtained for various possible combinations of depth of cut in finish and rough milling operation, to cut a total depth of cut of 5 mm.

$$(Z_{eq})_{total} = (Z_{eq})_{finish} + (Z_{eq})_{rough}. \qquad (26)$$

9. Evaluation of best solution

Table 9 shows various possible combinations of depth of cut in finish and rough milling operation. As shown in Table 9, the best solution is the one having minimum value of total equivalent objective function. It is observed from Table 9, that from production rate point of view, combination 1 {i.e. 1, 1, 1, 1, 1} is better, whereas, from cost point of view, combination 4 is better {i.e. 1, 2, 2}. The compromise best solution is 4 having minimum equivalent value (Z_{eq}) of 9.966.

Table 9. Possible combinations to cut a total depth of 5 mm

S.N.	afinish	arough	arough	arough	arough	Zeq	(T2)total	Utotal
1	1	1	1	1	1	10.314	3.309	5.647
2	1	1.5	1.5	1		10.130	3.409	5.259
3	1	2	1	1		10.140	3.377	5.331
4	1	2	2			9.966	3.445	5.015
5	1	2.5	1.5			9.993	3.454	5.027
6	1.5	2	1.5			10.124	3.585	5.105

Application of artificial bee colony (ABC) algorithm to solve the considered multi-objective multi-pass milling process leads to the following optimum solution.

- Total number of passes required = 3.
- Number of rough cutting passes = 2 with depth of cut in each pass = 2 mm.
- Number of finish pass = 1 with depth of cut in finishing pass = 1 mm.
- Feed per tooth for a = 2 mm in roughing operation = 0.154 mm/tooth
- Cutting speed for a = 2 mm in roughing operation = 52.91 m/min
- Feed per tooth for a = 1mm in finishing operation = 0.211 mm/tooth
- Cutting speed for a = 1 mm in finishing operation = 57.08 m/min
- Total production time = $T_{pr} = T_1 + T_2 = 2.550 + 3.445 = 5.995$ min
- Total production cost = $U + A_4 = 5.015 + 0.375 = \$ 5.39$

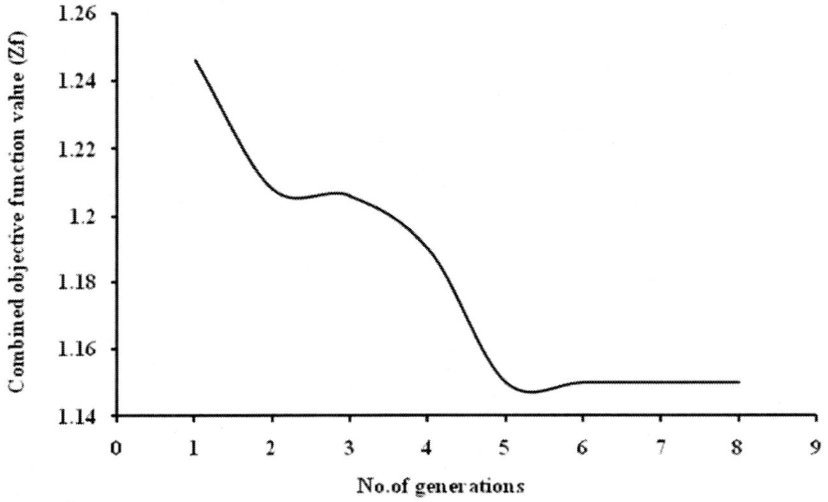

Figure 2. Convergence of ABC algorithm for finish milling.

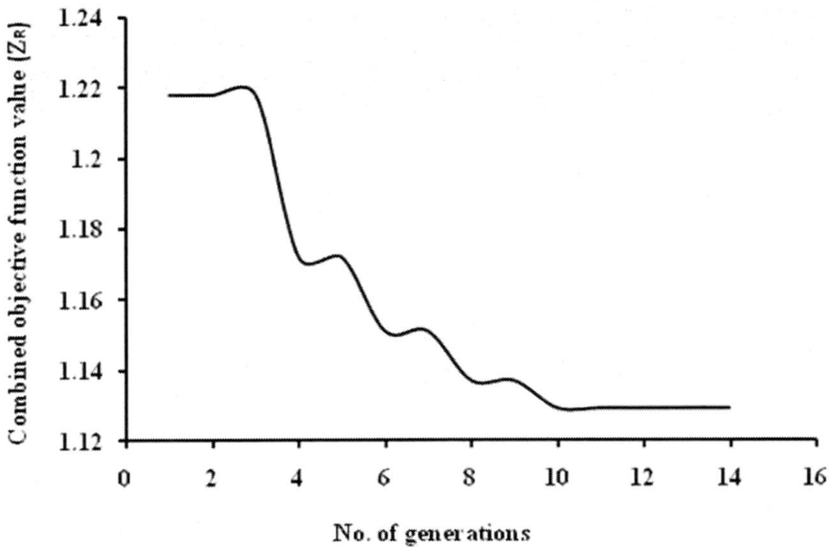

Figure 3. Convergence of ABC algorithm for rough milling

As shown in Figures 2 and 3, the convergence rate of artificial bee colony algorithm is very high and it requires little iteration for convergence to the optimal solution.

CONCLUSION

Multi-pass operations are often preferred to single pass operations for economic reasons. Determination of optimal process parameters such as the number of passes, depth of cut for each pass, speed, and feed is considered crucial in multi-pass machining. The effective optimization of the process parameters affects dramatically the cost and production time of machined components as well as the quality of the final products. In this chapter, multi-objective optimization aspects of multi-pass milling operation are considered. The two objectives considered are minimization of production time (i.e. maximization of production rate) and minimization of production cost subjected to the various practical constraints such as arbor strength, arbor deflection, cutting power, and surface roughness. Process parameters considered are feed per tooth, cutting speed, and depth of cut with their upper and lower bound values.

The performance of artificial bee colony (ABC) algorithm is studied in terms of convergence rate and accuracy of the solution. The algorithm is applied to obtain the optimum process parameter values for various selected cutting strategies. The optimum strategy is selected based on the compromise best of production time and production cost. The convergence rate of artificial bee colony algorithm is very high and it requires only few iterations for convergence to the optimal solution. The artificial bee colony algorithm can also be easily modified to suit optimization of process parameters of other machining processes such as grinding, turning, drilling, etc.

REFERENCES

[1] Karaboga, D. (2005). An idea based on honey bee swarm for numerical optimization. *Technical Report-TR06'*. Computer Engineering Department, Erciyes University.

[2] Karaboga, D, and Basturk, B. (2007). A powerful and efficient algorithm for numerical function optimization: artificial bee colony (ABC) algorithm. *Journal of Global Optimization, 39(3),* 459-471.

[3] Karaboga, D., and Basturk B. (2008). On the performance of artificial bee colony (ABC) algorithm. *Applied Soft Computing, 8*, 687-697.

[4] Shin, Y.C, and Joo, Y.S. (1992). Optimization of machining conditions with practical constraints. *International Journal of Production Research, 30*, 2907-2919.

[5] Wang, J. (1993). Multiple objective optimization of machining operations based on neural networks. *International Journal of Advanced Manufacturing Technology, 8*, 235-243.

[6] Tolouei-Rad, M., and Bidhendi I.M. (1997). On the optimization of machining parameters for milling operations. *International Journal of Machine Tools and Manufacture, 37*, 1–16.

[7] Sonmez, A.I., Baykasoglu, A., Dereli, T., and Filiz, I.H. (1999). Dynamic optimization of multi-pass milling operations via geometric programming. *International Journal of Machine Tools and Manufacture. 39*, 297–32.

[8] Agapiou, J.S. (1992). The optimization of machining operations based on a combined criterion – Part 2: Multi-pass operations. *Journal of Engineering for Industry, 114*, 508–513.

[9] Shunmugam, M.S., and Reddy. S.V.B. (2000). Narendran AA. Selection of optimal conditions in multi-pass face-milling using a genetic algorithm. *International Journal of Machine Tools and Manufacture, 40*, 401–414.

[10] Lui, Y.M., and Wang. C.J., (1999). A modified genetic algorithm based optimization of milling parameters. *International Journal of Advanced Manufacturing Technology, 15*, 796-809.

[11] Wang, Z.G., Rahman M., Wong, Y.S., and Sun, J. (2005). Optimization of multi-pass milling using parallel genetic algorithm and parallel genetic simulated annealing. *International Journal of Machine Tools and Manufacture, 45*, 1726–1734.

[12] Baskar, N., Asokan, P., Saravanan, R., and Prabhaharan, G. (2006). Selection of optimal machining parameters for multi-tool milling operations using a memetic algorithm. *Journal of Material Processing Technology, 174*, 239-249.

[13] Onwubolu, G.C. (2006). Performance-based optimization of multi-pass face milling operations using Tribes. *International Journal of Machine Tools and Manufacture, 46*, 717–727.

[14] Yildiz, A.R., (2009). A novel hybrid immune algorithm for optimization of machining parameters in milling operations. *Robotics and Computer-Integrated Manufacturing, 25(2)*, 261-270.

[15] Zarei, O, Fesanghary, M., Farshi, B., Saffar, R.J., and Razfar, M.R. (2009).Optimization of multi-pass face-milling via harmony search algorithm. *Journal of Materials Processing Technology, 209*, 2386-2392.

[16] Davim, J.P., and Antonio, C.A.C. (2001). Optimization of cutting conditions in machining of aluminium matrix composites using a numerical and experimental model. *Journal of Material Processing Technology, 112*, 78–82.

[17] Baskar, N., Saravanan, R., Asokan, P., and Prabhaharan, G. (2004). Ant colony algorithm approach for multi-objective optimization of surface grinding operations. *International Journal of Advanced Manufacturing Technology, 23*, 311–317.

[18] Saravanan, R., and Sachithanandam, M. (2001). Genetic algorithm (GA) for multivariable surface grinding process optimization using a multi-objective function model. *International Journal of Advanced Manufacturing Technology, 17(5)*, 330–338.

[19] Abburi, N.R. and Dixit, U.S. (2007). Multi-objective optimization of multipass turning processes. *International Journal of Advanced Manufacturing Technology, 32*, 902–910.

[20] Rao, R.V. (2007). *Decision making in the manufacturing environment using graph theory and fuzzy multiple attribute decision making methods.* London, Springer.

In: Artificial Intelligence in Manufacturing Research
Editor: J. Paulo Davim
ISBN 978-1-62618-557-9

Chapter 3

OPTIMIZATION OF ABRASIVE FLOW MACHINING PROCESS PARAMETERS USING PARTICLE SWARM OPTIMIZATION AND SIMULATED ANNEALING ALGORITHMS

***P. J. Pawar*[1], *R. Venkata Rao**[*1] *and J. P. Davim*[2]**

[1]Department of Mechanical Engineering,
S.V. National Institute of Technology, Ichchanath, Surat,
Gujarat, India
[2]Department of Mechanical Engineering, University of Aveiro,
Campus Santiago, Aveiro, Portugal

ABSTRACT

With the development of industry manufacturing technology, fine surface finish is in high demand in a wide spectrum of industrial applications. Abrasive flow machining (AFM) is an advanced finishing process capable of producing excellent surface finish of the order of few nanometers. AFM process is used to deburr, polish, radius, and remove recast layers of critical components in aerospace, automotive, electronic, and die-making industries. The AFM process is becoming popular due to its ability to give repeatable and consistent results. However, as the surface finish requirement increases, the operational cost of these processes increases exponentially. Quality, cost, time and efficiency of these processes can be improved significantly by choosing the optimum values of their process parameters.

This paper presents two advanced optimization algorithms known as particle swarm optimization (PSO) and simulated annealing (SA) to find the optimal combination of process parameters of AFM process. The results of the proposed algorithms are compared with the previously published results.

[*] Corresponding author, e-mail: ravipudirao@gmail.com

INTRODUCTION

Finishing operations represent a critical and expensive phase of overall manufacturing processes. The most labor intensive and uncontrollable area in the manufacturing of precision parts involves final finishing operations, which frequently demand as much as 15% of the total manufacturing cost. The dimensional and alignment accuracy and quality of surface finish are taken care of by finishing processes such as grinding, lapping, honing and super-finishing (i.e. traditional methods of finishing). But the applications of these traditional abrasive-finishing processes are limited to the production of work pieces of basic forms such as flat, cylindrical, etc. These finishing processes are being pushed to their limits of performance especially in components of hard materials, and complicated shapes. Also, the traditional fine finishing operations employ a rigid tool that subjects the work piece to substantial normal stresses, which may cause micro-cracks resulting in reduced strength and reliability of the machined part. Hence, there is a need to develop new finishing processes with wider bounds of application areas, better quality performance, higher productivity, and automatic operation. Abrasive flow machining (AFM) is such an advanced finishing process.

AFM is an advanced finishing process that can be used to deburr, radius, polish, remove recast layer, and to produce compressive residual stresses. This process was developed by the Extrude Hone Corporation, USA in 1960s as a method to deburr and polish difficult-to-reach surfaces and edges by flowing abrasive laden polymer with special rheological properties. However, as the use of the AFM process is a costly affair, the optimum selection of process parameters of these processes is essential for efficient and economic utilization of process capabilities. From the review of past literature, it is understood that very few efforts have been made for parameter optimization of AFM process. Thus, efforts must be extended by using advanced optimization algorithms which are more powerful, robust, and able to provide accurate solution. This paper is intended to apply two such optimization algorithms known as particle swarm optimization (PSO) and simulated annealing (SA) for optimization of process parameters of AFM process.

The next section presents optimization aspects of AFM process.

OPTIMIZATION ASPECTS OF ABRASIVE FLOW MACHINING PROCESS

AFM process removes small quantity of material by flowing a semisolid abrasive laden compound called 'media' (abrasive particles uniformly suspended in viscous chemical compound) through or across the surfaces of the work piece to be finished. As shown in Figure 1 [1], two vertically opposed cylinders extrude media back and forth through passages formed by the work piece and tooling. The machining action compares to a grinding or lapping operation as the media gently and uniformly abrades the surfaces and/or edges. The media acts as a 'self deformable stone' having protruding abrasive particles acting as cutting tools. The media is composed of semisolid carrier (e.g. polyborosilixane) and abrasive grains. The abrasive action during AFM depends on the extrusion pressure, flow volume and media flow speed determined by the machine setting in relation to media type, passage area, and media formulation which includes media viscosity, and abrasive type and size.

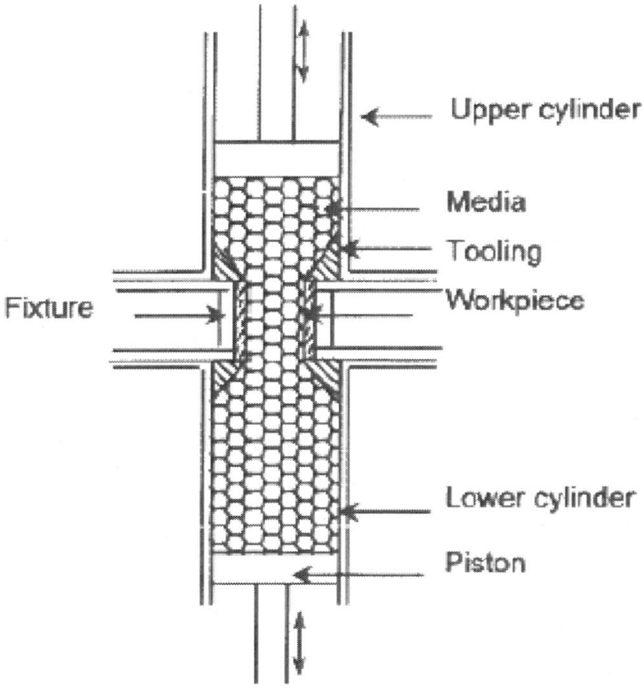

Figure 1. Schematic diagram of AFM process [1].

AFM process is gaining widespread attention due to its ability to produce consistent and predictable results. Removing stress raisers at sharp corners by producing controlled radii on edges can substantially improve thermal and mechanical fatigue strength of highly stressed components. Additional benefits over traditional finishing processes include a substantial time saving and better control with regard to the accuracy and squareness of the bearing surfaces. The process can deburr holes as small as 0.2 mm and radius edges from 0.025 - 1.5 mm. Tolerances can be held to ±5 μm [2,3].

Review of Literature on AFM Process

AFM process is relatively a new process and much information is not available in the literature that deals with optimization of the process parameters. The relationship between process parameters and performance characteristics are not known completely. The abrasion ability of abrasive media is governed by many factors, especially by grain size, abrasive concentration, extrusion pressure, and hardness of work piece material. In order to analyze the influence of such parameters and other AFM conditions upon material removal and surface roughness of the machined surface, experimental investigations were carried out by many researchers [1-4]. A stochastic modeling and analysis technique called data dependent system (DDS) was used by Williams and Rajurkar [4] to study the surface roughness profiles before and after AFM. Rajeshwar et al. [5] presented a simulation model to determine the characteristics of media flow during machining. The finite difference method was chosen for obtaining the solution.

Williams [6] used acoustic emission signals to analyze the mechanism of surface generation in AFM and compared them to the acoustic emission signals of grinding process. Jain and Adsul [1] studied the effects of different process parameters, such as number of cycles, concentration of abrasive, abrasive mesh size, and media flow speed, on material removal and surface finish in terms of polynomial equations.

Jain et al. [7] developed a back propagation neural network for modeling of AFM process. The inputs were media flow speed (v), percentage concentration of the abrasives (c), abrasive mesh size (d), and number of cycles (n). The outputs were material removal rate (MRR) and surface roughness (R_a). Based on the results of simulation, the possibility of using the neural network model for surface quality and material removal rate prediction for AFM process was confirmed. This model could be used to study the AFM process by examining the effects of the input process parameters on the performance of AFM process. Jain et al. [8] developed a model for the flow of AFM media through cylindrical work piece and solved the same by using finite element method (FEM). The model was shown to predict the radial stresses at the work piece surface with reasonable accuracy. The normal stress, so obtained from the flow model, was used for the estimation of material removal rate and surface roughness.

Centrifugal force assisted abrasive flow machining (CFAAFM) process has recently been tried as a hybrid machining process with the aim towards the performance improvement of AFM process by applying centrifugal force on the abrasive laden media with a rotating centrifugal force generating rod introduced in the work piece passage. Walia et al. [9] explored the application of centrifugal force for the productivity enhancement of the process. The authors had reported that centrifugal force enhances the material removal rate and improves the scatter of surface roughness value in AFM.

Ali-Tavoli et al. [10] proposed an approach using group method of data handling (GMDH)-type neural networks and genetic algorithms for modeling the effects of number of cycles and abrasive concentration on both material removal rate and surface roughness, using some experimentally obtained training and testing data for brass and aluminum in AFM process. However, the approach had not considered other important process input parameters such as media flow speed, abrasive mesh size, etc. Also, the approach is computationally more complex. Furthermore, genetic algorithms (GA) provide a near optimal solution for a complex problem having large number of parameters and constraints. This is mainly due to difficulty in determination of optimum controlling parameters.

Jain et al. [11] presented the details of process parameters optimization of AFM and magnetic abrasive finishing (MAF) processes using real-coded genetic algorithms. The authors described a surface roughness model that was developed to form the objective function for the optimization of AFM process. However, the results of GA presented by the authors are erroneous and ineffective.

Optimization Model of AFM Process

The optimization model for abrasive flow machining process is formulated in the present work based on the analysis given by Jain et al. [11]. The optimization model is formulated as shown below:

Objective function:

$$\text{Minimize } (R_a)_{Ns}=R_a^0[1-226814\{(K_{afm}/H_wR_a^0)(P_h/d)((1+\eta_a(A_{pmc}L_s/A_w)cdN_s)^{1/2}-1)\}^2]^2 \quad (1)$$

where, $(R_a)_{Ns}$ = Final surface value after N_s number of strokes; R_a^o = Initial surface roughness, H_w = Brinell hardness of work piece, P_h = Extrusion pressure, A_{pmc} = Cross-sectional area of piston of medium-containing cylinder, L_s = Stroke length of piston, A_w = Cross-sectional area of work piece, η_a = Proportion of abrasives effectively participating in machining, c = Volumetric concentration of abrasive particles, d = Mesh size, and K_{afm} = Proportionality constant relating normal radial stress acting on the abrasive grain.

Constraint: The constraint is to ensure that the final surface roughness value after AFM process is smaller than the initial surface roughness value. This is mathematically expressed as below:

$$1-226814\{(K_{afm}/H_wR_a^0)(P_h/d)((1+\eta_a(A_{pmc}L_s/A_w)cdN_s)^{1/2}-1)\}^2 \geq 0.0 \quad (2)$$

Decision variables (i.e. process parameters):

Four decision parameters were considered by Jain et al. [11] in the optimization problem and these were, concentration of abrasives by volume (c), abrasive mesh size (d), number of strokes (Ns), and extrusion pressure (Ph). The parameter bounds for these 4 decision parameters were as shown below.

$0.05 \leq c \leq 0.5$, $8 \leq d \leq 1000$, $1 \leq Ns \leq 100$, and $0.7 \leq Ph \leq 25$ MPa

The optimization model of AFM process is solved using PSO and SA algorithms which are described in the next two sections.

PARTICLE SWARM OPTIMIZATION

Particle swarm optimization (PSO) is an evolutionary computation technique developed by Kennedy and Eberhart [12]. It exhibits common evolutionary computation attributes including initialization with a population of random solutions and searching for optima by updating generations. Potential solutions, called particles, are then "flown" through the problem space by following the current optimum particles. The updates of the particles are accomplished as per the following equations.

$$V_{i+1}= w * V_i + c_1*r_1*(pBest_i - X_i) + c_2*r_2*(gBest_i - X_i) \quad (3)$$

$$X_{i+1} = X_i + V_{i+1} \quad (4)$$

Equation (3) calculates a new velocity (V_{i+1}) for each particle (potential solution) based on its previous velocity, the best location it has achieved (‘*pBest*’) so far, and the global best location ('*gBest*'), the population has achieved. Equation (4) updates individual particle's position (X_i) in solution hyperspace. The two random numbers 'r_1' and 'r_2' in equation (3) are independently generated in the range [1].

The acceleration constants 'c_1' and 'c_2' in equation (3) represent the weighting of the stochastic acceleration terms that pull each particle towards '*pBest*' and '*gBest*' positions. 'c_1' represents the confidence the particle has in itself (cognitive parameter) and 'c_2' represents the confidence the particle has in swarm (social parameter). Thus, adjustment of these constants changes the amount of tension in the system. Low values of the constants allow particles to roam far from target regions before being tugged back, while high values result in abrupt movement toward, or past through target regions [13]. The inertia weight 'w' plays an important role in the PSO convergence behavior since it is employed to control the exploration abilities of the swarm. The large inertia weights allow wide velocity updates allowing to globally explore the design space while small inertia weights concentrate the velocity updates to nearby regions of the design space. The optimum use of the inertia weight "w" provides improved performance in a number of applications.

Unlike genetic algorithm, PSO algorithm does not need complex encoding and decoding process and special genetic operator. PSO takes real number as a particle in the aspect of representation solution and the particles update themselves with internal velocity. In this algorithm, the evolution looks only for the best solution and all particles tend to converge to the best solution. In the implementation process, particles randomly generated at the beginning or generated by internal velocity during the evolutionary process usually violate the system constraints resulting in infeasible particles. Therefore, the handling of system constraints, particularly nonlinear equation constraints, and the measurement and evaluation of infeasible particles is very important. To cope with constrained problems with evolutionary computation, various approaches such as rejection of infeasible individuals, repair of infeasible individuals, replacement of individuals by their repaired versions, and penalty function methods can be adopted. Among them, the penalty function method is particularly promising as evidenced by recent developments [13] and the same is adapted in the present work.

SIMULATED ANNEALING

Simulated annealing is a probabilistic hill climbing soft computing algorithm. A probabilistic algorithm is one in which the objective function is evaluated at sample of points chosen randomly from the feasible region. A hill climbing technique is one, which allows for an increase in the value of objective function in a controlled manner. The methodology of simulated annealing algorithm is described below.

Let a feasible configuration (k) is a point in an allowable region, T is a controlling parameter and the cost C(k) is the quantity to be minimized. The neighborhood of a configuration is a set of predefined feasible points from which next configuration is chosen. If T is a quantity analogous to the temperature in annealing of solids, then the manner in which the temperature is to be decremented and the number of moves required is to be decided. The cost function is said to be in the state of equilibrium at temperature T when probability of being in the configuration k is:

$$\Pr\{config. = k\} = \frac{e^{-C(k)/T}}{Z(T)} \tag{5}$$

where $Z(T) = \sum_{j} e^{-C(k)/T}$, (6)

If ‘i’ is the current configuration with cost C(i) then using the Metropolis algorithm [14], we can say that the probability of accepting ‘j’ as next configuration depends on the difference in the function value at these two points or on $\Delta C = C(j) - C(i)$ and is calculated using the Boltzman probability distribution:

$$\Pr\{new = j \mid current = i\} = \begin{cases} 1 & if\ \Delta C \leq 0 \\ e^{-\Delta C/T} & otherwise \end{cases} \tag{7}$$

Thus the probability of accepting new configuration is either 1 or $e^{-\Delta C/T}$ depending upon the sign of ΔC .

The nest section provides the application of proposed PSO and SA algorithms.

APPLICATION OF PSO AND SA ALGORITHMS TO AFM PROCESS PARAMETERS OPTIMIZATION

The optimization of process parameters of AFM process using the proposed PSO and SA algorithms is discussed below based on the model given by Jain et al. [11].

Optimization Using PSO Algorithm

To ensure the convergence of PSO algorithm, the condition specified by equation (8) must be satisfied [15].

$$\max(|\lambda_1|, |\lambda_2|) < 1 \tag{8}$$

where, λ_1 and λ_2 are the eigen values given by equations (13) and (14).

$$\lambda_1 = (1 + w - \phi_1 - \phi_2 + \gamma) / 2 \tag{9}$$

$$\lambda_2 = (1 + w - \phi_1 - \phi_2 - \gamma) / 2 \tag{10}$$

and $\gamma = [(1 + w - \phi_1 - \phi_2)^2 - 4w)]^{1/2}$ (11)

$\phi_1 = r_1 * c_1$ and $\phi_2 = r_2 * c_2$

Considering the feasible range for the value of '$\phi_1 + \phi_2$' as 0 to 4 and that for 'w' as 0 to 1, it can be observed that for convergent trajectories the relation given by equation (12) must be satisfied.

$$w > 0.5\,(\phi_1 + \phi_2) - 1 \tag{12}$$

Now, in the present study the following values of 'w', 'c_1' and 'c_2' are used.

- Inertia weight factor (w) = 0.65
- Acceleration coefficients: $c_1 = 1.65$ and $c_2 = 1.75$

Considering the extreme possibility of random number as 'r_1'=0.95 and 'r_2'=0.95, the right hand term in equation (16) is 0.5*(0.95*1.65 + 0.95*1.75) –1 = 0.61, which is less than 0.65 thus satisfies the equation (16). Hence, the values of 'w', 'c_1' and 'c_2' selected in the present work are appropriate for convergence of the algorithm.

In the present work values of the constants shown in equations (1) and (2) are considered as $A_{pmc} = 5026.5$ mm^2, $A_w = 484$ mm^2, $L_s = 80$ mm, $H_w = 2000$ MPa, $\eta_a = 0.2$, $K_{afm} = 0.4$, and $R_a^o = 3.0$ μm.. For the selected values of coefficients 'c_1', 'c_2' and inertia weight 'w' as discussed above, the convergence of the PSO algorithm is shown in Figure 2.

The optimum process parameter values obtained by using PSO algorithm are given as:

c = 0.32 (i.e. 32%), d = 1000, $P_h = 15.96$ N/mm^2, $N_s = 73.10$ (~73)

Optimum value of final surface roughness = 0.000026 μm

Value of surface finish improvement constraint (given by equation (2)) = 0.00293

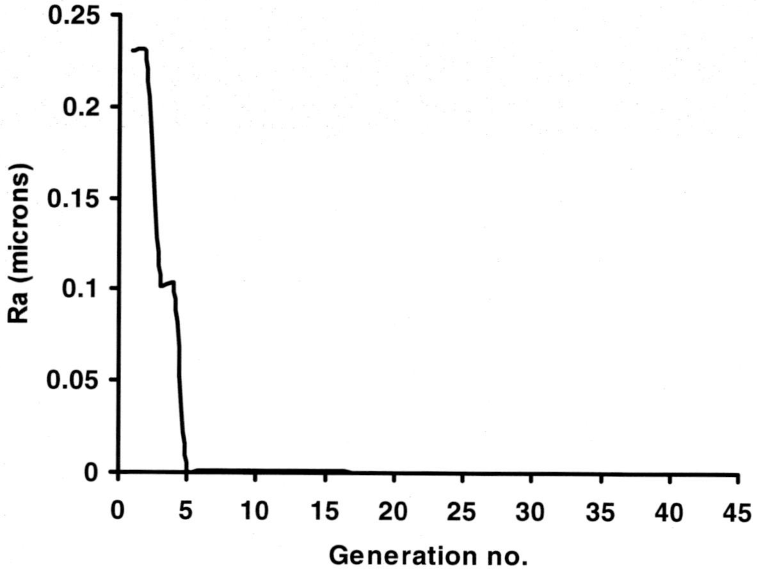

Figure 2. Convergence of PSO algorithm for AFM process.

The results obtained by using PSO algorithm showed that the final roughness value is almost equal to zero (i.e. 0.000026 μm) without violating the surface roughness constraint. Thus, the results obtained by PSO algorithm provide large improvement in the solution. Table 1 shows comparison of the results of optimization using GA, PSO, and SA algorithms.

Validity of the model and the optimality of the above mentioned solution could be confirmed from the Figures 3 to 4. As shown in Figure 3, the surface roughness initially decreases up to a certain value and then increases with increase in volumetric concentration of abrasives. This nature is confirmed experimentally by Jain and Adsul [1]. Also the surface finish improvement constraint continuously decreases with increase in concentration of abrasives. For the optimization problem under consideration, the surface finish decreases up to value of ‘c’= 0.32 (i.e.32%) after which surface roughness increases as shown in Figure 3. Also the constraint value is violated for any value of ‘c’ more than 0.32. Thus the selection of optimum value of ‘c’ = 0.32 is appropriate.

Table 1. Comparison of optimization results of AFM process

Optimization technique	c	d	Ns	Ph	Ra	Constraint value
GA [11]	0.196	851	23.3	0.81	0.1041a	0.18628a
GA	0.196	851	23.3	0.81	2.996b	0.999b
PSO	0.32	1000	73.10	15.96	0.000026	0.00293
SA	0.217	622.3	73.69	15.20	0.000105	0.00705

a: Values wrongly calculated by Jain et al. [11].
b: corrected values.

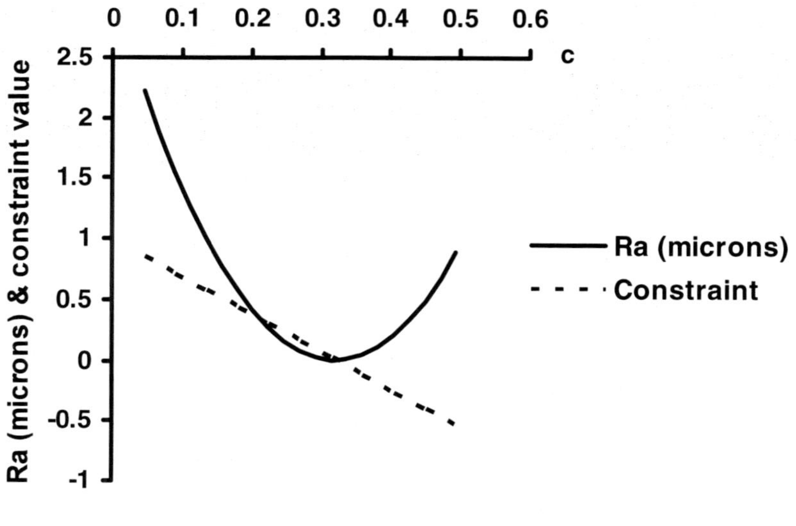

Figure 3. Variation of surface roughness and constraint with volumetric concentration of abrasive particle ‘c’.

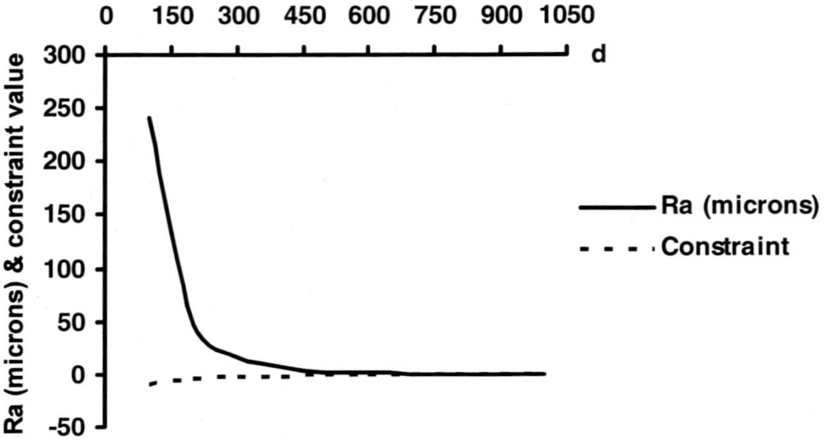

Figure 4. Variation of surface roughness and constraint with abrasive mesh size 'd'.

Figure 4 shows the variation of surface roughness and constraint with abrasive mesh size. The surface roughness decreases with increase in 'd' following the law of diminishing return. Also the constraint is violated for lower values of 'd'. Hence the upper bound value of 'd' = 1000 is selected as an optimum value.

Figure 5 shows variation of surface roughness and constraint with number of strokes. As shown in Figure 5, the surface roughness initially decreases up to a certain value and then increases with increase in number of strokes. Also the surface finish improvement constraint continuously decreases with increase in concentration of abrasives. For the optimization problem under consideration, the surface roughness decreases up to value of 'N_s'= 73.10 after which surface roughness increases as shown in Figure 5. The constraint is also getting violated after this value of 'N_s'. Hence the optimum value of number of strokes 'Ns' = 73.10 is selected.

Figure 6 shows variation of surface roughness and constraint with extrusion pressure. As shown in Figure 6, minimum value of surface roughness occurs for extrusion pressure 'P_h' = 15.96 without violating constraint surface roughness improvement constraint. Hence the optimum value of 'P_h'=15.96 MPa is selected.

It is also observed from the numerical results that very high surface finish of 0.000026 μ is achieved by selecting optimum parameters as provided by PSO. Although this value is not practically attainable as we cannot machine the material below its atomic size, but it can be ensured that with the given set of parameters, maximum possible and attainable value of surface finish will be obtained, as the values of suggested optimum parameters lies within their specified bounds and also the set of optimum parameters satisfies the constraint on surface roughness improvement.

Keeping in view of the above discussion, one should consider the limiting but practically infeasible case of surface improvement constraint value being equal to zero. The theory of constrained optimization mentions that a feasible solution (lying in the feasible region) that optimizes the objective function and gives smaller value of the constraint is likely to be an optimum solution. Therefore, those values of the decision variables lying in the feasible region and minimizing the final surface roughness value and surface finish improvement constraint should be the optimum solution.

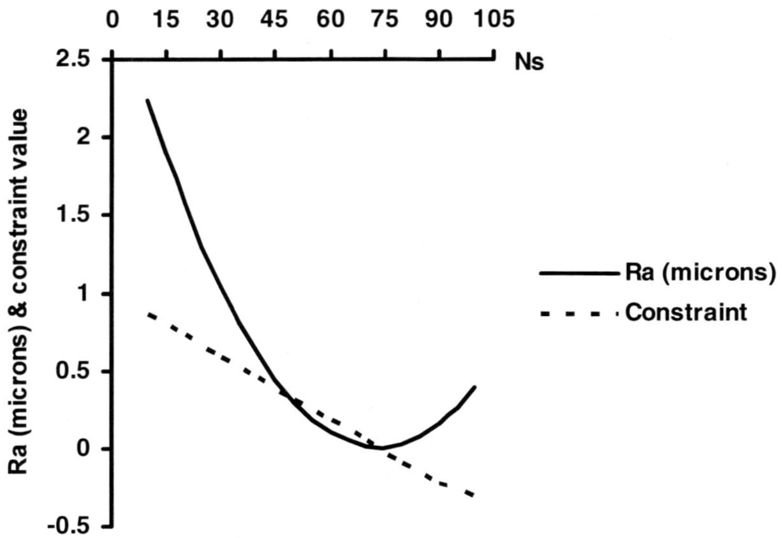

Figure 5. Variation of surface roughness and constraint with number of strokes ‘Ns’.

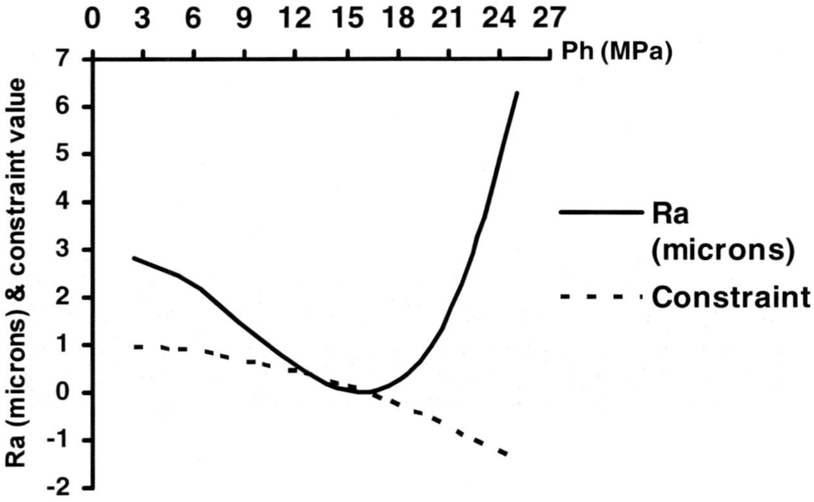

Figure 6. Variation of surface roughness and constraint with extrusion pressure (Ph).

Table 2 presents the results of optimization of process parameters of AFM for various parameter values of w, c_1 and c_2. It can be observed that the feasible values of surface roughness and the surface finish improvement constraint are 0.0785 and 0.1607 respectively corresponding to w = 0.45, c_1 = 1.55 and c_2 = 1.45 and the corresponding feasible values for c, d, N_s, and P_h are 0.27, 500, 55 and 13 respectively. These feasible values are better than those presented by Jain et al.[11] who suggested the feasible values of surface roughness and the surface finish improvement constraint as are 0.1041 and 0.18628 respectively. Thus PSO results of optimization are proved better than the GA results given by Jain et al. [11].

Table 2. Results of optimization of AFM process for different values of w, c_1, c_2 of PSO

S.No	*w*	*c1*	*c2*	*c*	*d*	*Ns*	*Ph*	*Ra*	Constraint
1	0.20	1.15	1.15	0.302	476.38	48.78	12.08	0.180	0.245
2	0.25	1.10	1.10	0.294	416.17	48.58	12.39	0.0426	0.119
3	0.30	1	1	0.305	433.47	53.07	12.36	0.0064	0.0462
4	0.45	1.55	1.45	0.27	500	55	13	0.0785	0.1607
5	0.65	1.65	1.75	0.32	1000	73.1	15.96	0.000026	0.0029
6	0.85	1.95	2	0.50	486	33.2	12.92	0.00362	0.0347
7	0.95	1.90	1.9	0.30	1000	30.71	25	0.0041	0.037

c = Concentration of abrasives by volume, d = Abrasive mesh size, Ns = Number of strokes, Ph=Extrusion pressure, and
Ra = Surface roughness value.

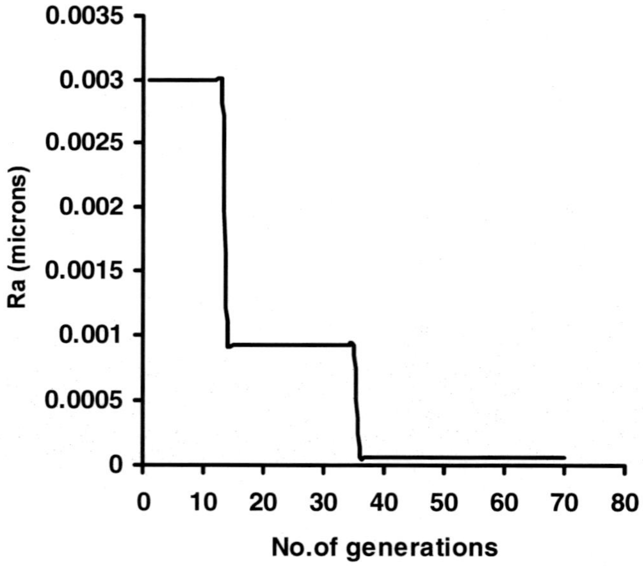

Figure 7. Convergence of SA algorithm for optimization of AFM process.

Optimization Using SA Algorithm

Now using the simulated annealing technique, the objective function considering maximization of the difference between the initial and final roughness value is written as:

$$Min.\ Z = Z_1 - Penalty * Z_2 \tag{13}$$

where, Z = Combined objective function.

Z_1 = Objective function given by equation (14)

Z_2 = Constraint given by equation (15)

For the present case the penalty is considered as 1 if $Z_2 < 0$; else penalty = 0.

The initial temperature is obtained by calculating the average of the function values at a boundary points. The convergence of the SA algorithm for optimization of AFM process parameters is shown in Figure 7.

As shown in Table 1, the results of genetic algorithm [11] are erroneous and the corrected results are ineffective. The results obtained by using SA algorithm are thus better than the results obtained by using GA. However, the results obtained by using PSO algorithm are better than the results provided by both GA and SA.

CONCLUSIONS

Abrasive flow machining (AFM) process is an economic and effective advanced finishing process which is capable of providing excellent surface finish on difficult to approach regions on a wide range of components. With AFM process, it has become possible to substitute various time-consuming deburring and polishing operations that had often led to non-reproducible results. In the present work, optimization aspects of process parameters of AFM process are considered using particle swarm optimization (PSO) and simulated annealing (SA) algorithms. The objective considered is, minimization of surface roughness subjected to a constraint and parameter bounds.

The comparison of the performance of various algorithms in terms of accuracy of solution and rate of convergence reveals that PSO algorithm outperformed the other algorithms such as genetic algorithm and simulated annealing. The results obtained by using PSO algorithm showed significant improvement over other optimization techniques such as genetic algorithm (GA) and simulated annealing (SA). On the other hand, the results of SA are better than the results of GA. Compared to other advanced optimization methods, few trials are required to predict the best and worst operating parameters of PSO algorithm and the algorithm requires only 30 to 40 iterations for convergence to the optimal solution. Unlike genetic algorithm, PSO algorithm does not need complex encoding and decoding process and special genetic operator. The proposed PSO and SA algorithms can be easily modified to suit single or multi-objective optimization of process parameters of other advanced finishing processes such as magnetic abrasive finishing, magnetic float polishing, ion beam machining, etc.

REFERENCES

[1] Jain, V.K., Adsul, S.G. (2000). Experimental investigation into abrasive flow machining. *International Journal of Machine Tools and Manufacture*, *40*, 1003-1021.

[2] Rhoades, L.J. (1987). Abrasive flow machining with not-so-silly putty. *Metal Finishing*, July 27-29.

[3] Rhoades, L.J. (1991). Abrasive flow machining: a case study. *Journal of Materials Processing Technology*, *28,* 107-116.

[4] Williams, R.E., Rajurkar, K.P. (1992). Stochastic modeling and analysis of abrasive flow machining. Transactions of ASME, Journal of Engineering for Industry, *114*, 74-81.

[5] Rajeshwar, G., Kozak, J., Rajurkar, K.P. (1994). Modeling and computer simulation of medium flow in abrasive flow machining process. *Proceedings of the International Mechanical Engineering Congress and Exposition*, Chicago, 965-971.

[6] Williams, R.E. (1998). Acoustic emission characteristics of abrasive flow machining. *Transactions of ASME, Journal of Manufacturing Science and Engineering*, 120, 264-271.

[7] Jain, R.K., Jain, V.K., Kalra, P.K. (1999). Modeling of abrasive flow machining process: a neural network approach. *Wear*, *231*, 242-248.

[8] Jain, R.K., Jain, V.K., Dixit, P.M. (1999). Modeling of material removal and surface roughness in abrasive flow machining process. *International Journal of Machine Tools and Manufacture*, *39*, 1903-1923.

[9] Walia, R.S., Shan, H.S., Kumar, P. (2006). Multi-response optimization of CFAAFM process through Taguchi method and utility concept. *Materials and Manufacturing Processes, 21*, 907-914.

[10] Ali-Tavoli, M., Zadeh, N.N., Khakhali, A., Mehran, M. (2006). Multi-objective optimization of abrasive flow machining processes using polynomial neural networks and genetic algorithms. *Machining Science and Technology*, *10*, 1-20.

[11] Jain, N.K., Jain, V.K., and Jha, S. (2007). Parametric optimization of advanced fine-finishing processes. *International Journal of Advanced Manufacturing Technology*, *34*, 1191-1213.

[12] Kennedy, J., and Eberhart, R. (1995). Particle swarm optimization. *Proceedings of IEEE, International Conference on Neural Networks, 4*, 1942-1948.

[13] Dong, Y., Tang, J., Xu, B., and Wang, D. (2005). An application of swarm optimization to nonlinear programming. *Computer and Mathematics with Applications*, *49*, 1655-1668.

[14] Metropolis, N., Rosenbluth, A., Rosenbluth, M., Teller, A., and Teller, E. (1953). Equation of state calculations by fast computing machines. *Journal of Chemical Physics*, *21*, 1087-1092.

[15] Bergh, F., and Engelbrecht, A.P. (2006). A study of particle swarm optimization particle trajectories. *Information Sciences*, *176*, 937-971.

In: Artificial Intelligence in Manufacturing Research
Editor: J. Paulo Davim
ISBN 978-1-62618-557-9

Chapter 4

STUDY OF EFFECTS OF PROCESS PARAMETERS ON BURR HEIGHT IN DRILLING OF AISI 316 STAINLESS STEEL USING ARTIFICIAL NEURAL NETWORK MODEL

V. N. Gaitonde[1,a], S. R. Karnik[b] and J. Paulo Davim[c]

[a]Department of Industrial and Production Engineering,
B.V.B. College of Engineering and Technology,
Hubli, India

[b]Department of Electrical and Electronics Engineering,
B.V.B. College of Engineering and Technology,
Hubli, India

[c]Department of Mechanical Engineering,
University of Aveiro, Campus
Santiago, Aveiro, Portugal

ABSTRACT

Burr size at the exit of holes in drilling degrades the precision of products and must be removed in subsequent deburring processes to allow the components to meet specified tolerances. Therefore understanding the drilling burr formation and its process parameters is essential for predicting the burr size, which in turn provides useful information to minimize the burr formation. An artificial neural network (ANN) based model has been developed to study the effect of process parameters such as cutting speed, feed, drill diameter, point angle and lip clearance angle on burr height during drilling of AISI 316 stainless steel. A multilayer feed forward ANN; trained using error back-propagation training algorithm (EBPTA) has been employed. The input-output patterns required for ANN training are obtained from the drilling experiments planned through the Box-Behnken design. The simulation results demonstrate the effectiveness of ANN based model to analyze the effects of drilling process parameters on burr height.

[1] Corresponding author, E-mail: gaitondevn@yahoo.co.in

Keywords: Drillig; AISI 316 stainless steel; Burr height; Box-Behnken design; Artificial neural network.

1. Introduction

The burr is a plastically deformed material generated on the edge of the drilled component. These burrs cause several problems for the quality and functionability of the products. There are no tools available for deburring if the exit burr is formed inside the cavity. Hence, special tools must be used, which in turn increases the deburring cost. Therefore, it is essential to understand the factors affecting the burr formation at the exit of holes in drilling so as to reduce the burr at the manufacturing stage. This necessitates accurate models to analyze the effects of drilling process parameters on burr height.

The finite element models, empirical drilling charts and the response surface methodology (RSM) based models were employed in the past to study the influence of various parameters on burr formation during drilling of several workpiece materials [1, 2]. The AISI 316 stainless steel work material finds broad applications in chemical industries, aircraft designs and in the manufacture of medical apparatus due to high corrosive resistance properties. But, the burr formation in AISI 316 stainless steel drilling is a greater concern due to high ductility of the work material. Hence, an attempt has been made in this work to analyze the interaction effects of process parameters such as cutting speed, feed, drill diameter, point angle and lip clearance angle on exit burr height during drilling. A multi-layer feed forward artificial neural network (ANN) has been employed to capture the relationship between the burr height and the affecting process parameters. The experiments required for ANN training were planned as per Box-Behnken design (BBD) during drilling of AISI 316 stainless steel workpieces using HSS twist drills.

2. Artificial Neural Network Modeling

The artificial neural network (ANN) is a flexible modeling tool with an aptitude to learn the mapping between input and output process parameters [3]. The purpose of ANN development is to imitate human brain so as to implement the various functions such as association, self-organization and generalization. The ANNs are parallel computer models of processes and the mechanisms, which constitute biological nerve systems. The ANNs are attractive in view of their high execution speed and modest computer hardware requirements in addition to an adaptive nature and capability in solving complex and non-linear problems. The ANN is made up of neurons connected via links. The information is processed within a neuron and is propagated to other neurons through the links connecting neurons.

In this investigation, a multi-layer feed forward ANN using error back propagation training algorithm (EBPTA) has been adapted. The EBPTA is a supervised learning algorithm based on generalized delta rule [3], which requires a set of inputs and desired outputs, known as training patterns. The EBPTA uses a gradient search technique, which updates the synaptic weights during learning stage in such a way that mean square error between the actual output pattern and the desired output pattern of network is minimized. The multilayer feed forward

ANN consists of neurons divided into input layer, hidden layers and the output layer. The net activation input for i^{th} neuron is given by [3]:

$$net_i = \sum_{j=1}^{n} w_{ij} x_j \tag{1}$$

where, w_{ij} = weight of link connecting neuron i to j; x_j = the output of jth neuron . For an unipolar sigmoid transfer function, the ouput of ith neuron is given as [3]:

$$o_i = \frac{1}{1+e^{\eta\, net_i}} \tag{2}$$

where, η is the scaling factor. The training algorithm used here is based on the weight updates so as to minimize the sum of squared error for K number of neurons in the output layer and is given by [3]:

$$E = \frac{1}{2} \sum_{k=1}^{K} (d_{k,p} - o_{k,p})^2 \tag{3}$$

where, $d_{k,p}$ = desired output for p^{th} pattern. The synaptic weights of the links are updated as [3]:

$$w_{ji(n+1)} = w_{ji(n)} + \alpha\, \delta_{pj} o_{pi} + \beta\, \Delta w_{ji(n)} \tag{4}$$

where, n is the learning step, α is the learning rate and β is the momentum constant. The error term δ_{pj} in Eq. 4 is given by [3]:

For output layer: $\delta_{pk} = (d_{kp} - o_{kp})(1 - o_{kp})$; $k=1, \ldots K$ (5)

For hidden layer: $\delta_{pj} = o_{pj}(1 - o_{pj}) \Sigma \delta_{pk} w_{kj}$; $j = 1, \ldots J$ (6)

where, J is the number of neurons in the hidden layer.

The steps involved in ANN training using EBPTA are mentioned below:

(1) The network weights are initialized to small random values.
(2) The input/desired output pairs are presented one by one, updating the weights each time.
(3) The mean square error (MSE) due to all outputs and NP number of patterns is computed as, $MSE = \frac{1}{NP} \sum_{p=1}^{NP} \sum_{k=1}^{K} (d_{kp} - o_{kp})^2$ (7)
(4) If (MSE < specified tolerance) or (epochs > $(\text{epochs})_{max}$)
Then stop.
Else, goto step 2.

3. Experimental Details

The database required for the development of ANN is obtained through the drilling experiments. As, the classical experimental design methods are too complex, an experimental layout plan based on Box-Behnken design (BBD) [4] has been selected to perform the experiments. In the current study, cutting speed, feed, drill diameter, point angle and lip clearance angle are selected as the process parameters. Three levels for each of the process parameters are identified. Usually, in industries during the manufacture of valves for the chemical plants, the drilling of stainless steel material is performed using HSS twist drills with the cutting speed in the range 8 - 16 m/min. The feed in the range 0.04 - 0.12 mm/rev is generally preferred with higher drill diameters in order to avoid excessive temperature rise during machining. The range of point angle for drilling of stainless steel was selected based on the investigations carried out by Stein [1]. The levels of lip clearance angle were identified through the preliminary experiments. Accordingly, the ranges of the process parameters were selected in the current study. The experimental parameters and their levels identified are presented in Table 1. Table 2 illustrates the experimental layout plan as per BBD, which consists of 46 sets of process parameter combinations.

A three-axis '*YCM-V116B*' CNC vertical machining center (Make: Yeong Chin Machinery Industries Co., Taiwan) was used to conduct the drilling experiments. The machining center is equipped with a maximum feed of 5000 mm/min and a variable spindle speed from 45 - 4000 rpm with a 15 kW drive motor. The maximum table travel along X – axis is 1100 mm, along Z – axis is 630 mm and the maximum saddle travel along Y – axis is 600 mm. The drilling fixture was used to clamp the specimens on a flat surface and the fixture was mounted in the vise on the table of machining center.

The AISI 316 stainless workpieces of 320 mm × 60 mm × 25 mm were used for all the drilling experiments. The chemical composition and mechanical properties of work material are listed in Table 3. The workpieces were polished on exit surface before drilling for the preparation of burr height measurements. The HSS parallel shank stub series twist drills (Make: Addison and Co. Ltd., India) confirming to IS: 5100/DIN: 1897/BS: 328/ ISO specifications were utilized for drilling experiments. The required point angle and lip clearance angles were ground as per BBD. '*Cut60EP*' water-soluble oil was used as coolant through out the experimentation.

Table 1. Experimental parameters and their levels

Parameter	Notation	Unit	Levels		
			1	2	3
Cutting speed	v	m/min	8	12	16
Feed	f	mm/rev	0.04	0.08	0.12
Drill diameter	d	mm	16	22	28
Point angle	θ	degree	118	126	134
Lip clearance angle	ψ	degree	8	9	10

The burr height was measured on '*RPP-400*' toolmakers' microscope (Make: Sicherun-Gen Versehen, Germany) with a resolution of 1 micron at 30× magnification. To measure the burr height, the focus was put on top of burr and then on exit surface. The burr height is the distance between two foci. It was observed that the burrs were more or less uniform and hence the burr height values were recorded at four equally spaced locations around the circumference and the average reading was taken as the process response. The experimental layout plan along with the observed values of burr height is given in Table 2.

4. RESULTS AND DISCUSSION

4.1. ANN Training

The experimental results as shown in Table 2 were utilized for the development of ANN model. The training of ANN for 46 input-output cominations has been performed using neural network tool box avilable in '*MATLAB*' software [4].

Table 2. Experimental layout plan and measured values of burr height

Trial no.	Process parameter settings					Burr height (mm)
	v (m/min)	f (mm/rev)	d (mm)	θ (degree)	ψ (degree)	
1	8	0.08	22	126	8	0.056
2	16	0.08	22	126	8	0.276
3	12	0.12	28	126	9	0.752
4	12	0.04	28	126	9	0.132
5	12	0.08	22	134	10	0.341
6	12	0.08	22	126	9	0.842
7	16	0.04	22	126	9	0.818
8	12	0.08	22	126	9	0.842
9	12	0.08	28	134	9	0.016
10	12	0.08	22	118	8	0.009
11	8	0.08	22	134	9	0.022
12	12	0.04	22	118	9	0.025
13	16	0.08	16	126	9	1.113
14	16	0.08	22	134	9	0.578
15	12	0.08	28	126	8	0.010
16	12	0.08	22	134	8	0.022
17	12	0.08	22	126	9	0.842
18	12	0.04	22	126	8	0.147
19	12	0.12	22	134	9	0.370
20	12	0.08	16	126	10	1.179
21	12	0.12	22	118	9	0.331
22	16	0.12	22	126	9	1.322
23	12	0.04	16	126	9	1.227
24	12	0.08	22	126	9	0.842
25	12	0.08	16	134	9	0.181
26	8	0.08	16	126	9	0.610
27	12	0.12	16	126	9	1.024

Table 2. (Continued)

Trial no.	Process parameter settings					Burr height (mm)
	v (m/min)	f (mm/rev)	d (mm)	θ (degree)	ψ (degree)	
28	12	0.08	16	118	9	0.419
29	12	0.12	22	126	10	0.895
30	12	0.08	22	126	9	0.842
31	8	0.08	22	118	9	0.012
32	12	0.04	22	126	10	1.114
33	12	0.12	22	126	8	0.755
34	12	0.08	16	126	8	0.354
35	8	0.08	22	126	10	0.371
36	8	0.08	28	126	9	0.007
37	8	0.04	22	126	9	0.595
38	16	0.08	28	126	9	0.376
39	12	0.08	22	126	9	0.842
40	12	0.08	28	118	9	0.009
41	12	0.08	28	126	10	0.219
42	16	0.08	22	118	9	0.014
43	8	0.12	22	126	9	0.541
44	12	0.08	22	118	10	0.013
45	12	0.04	22	134	9	0.377
46	16	0.08	22	126	10	1.072

Table 3. Chemical composition and mechanical properties of AISI 316 stainless steel

Chemical composition (wt %)	0.08 C, 2.5 Mo, 2.0 Mn, 1.0 Si, 0.03 S, 0.045 P, 12.0 Ni, 17.0 Cr, 65.0 Fe
Mechanical properties	Tensile strength: 620 MPa Yield strength: 415 MPa Hardness: 190 BHN Modulus of elasticity: 193 GPa

All the inputs and the desired outputs were normalized using the following equation:

$$X_{normal} = \frac{2(X - X_{min})}{(X_{max} - X_{min})} - 1 \tag{8}$$

where, X_{min} = minimum value in the matrix of pattern for X; X_{max} = maximum value in the matrix of pattern for X. This normalization maps all the inputs and desired outputs between –1 and +1. The simulated multi-layer feed forward ANN architecture consists of 5 neurons in the input layer (corresponding to five input parameters, v, f, d, θ and ψ), 1 neuron in the output layer (corresponding to one output parameter). The ANN training simulation was performed using a variable learning rate training procedure "*traingdx*" of MATLAB NN toolbox [5]. A trial and error procedure has been employed to optimize the number neurons in the hidden layer. Appropriate learning rate parameters for faster convergence and the momentum factor for increasing the rate of learning have been employed. The training has been continued until the mean squared error (MSE) reaches 10^{-5} or 5000 epochs. After successful training, one hidden layer with 11 neurons in the ANN structure with a learning rate of 0.3 and a momentum constant of 0.5 were found to be suitable for burr height model.

4.2. ANN Testing

The ANN was initially tested with 46 input training patterns. For each input training pattern, the ANN predicted burr height value was compared with the respective experimental measured value and were found to be very close for each of the training patterns. For the validation purpose, the drilling experiments were performed for 12 new combinations of input process parameters, which do not belong to the training data set. The experimental conditions along with the measured burr height values are summarized in Table 4. The comparison of the ANN predicted and the measured values of burr height for the validation data set is depicted in Figure 1. It can be seen that the ANN predicted values follow almost the same trend as that of the measured values, which validates the developed ANN burr height model.

Table 4. Experimental conditions and measured values of burr height for validation data

Test trial no.	Levels of process parameters					Burr height (mm)
	v (m/min)	f (mm/rev)	d (mm)	θ (degree)	ψ (degree)	
1	8	0.04	22	134	10	0.781
2	8	0.04	28	134	10	0.053
3	8	0.04	16	134	10	1.112
4	8	0.04	22	134	10	0.773
5	12	0.12	28	126	10	0.612
6	12	0.12	22	134	8	0.187
7	12	0.12	22	134	10	0.681
8	16	0.12	22	126	8	0.813
9	16	0.12	22	126	10	1.834
10	16	0.12	16	126	10	1.221
11	16	0.04	16	126	9	1.323
12	16	0.04	16	134	9	0.421

4.3. Parametric Analysis of Process Parameters on Burr Height

The developed ANN model was utilized to study the interaction effects of selected process parameters on burr height. To analyze the interaction effects, the 3D response surface plots were generated considering two parameters at a time, while the other parameters are held constant at their respective center levels. These interaction plots are presented in Figures

2-11. It is observed from Figure 2 that for a given cutting speed, the burr height increases with increase in feed. A similar trend is observed with the cutting speed variations for a specified feed. However, it is to be noted that, the burr height is highly sensitive at higher values of cutting speed and feed. Moreover, minimum burr height results with lower values of cutting speed and feed combination. From the interaction plot of cutting speed and drill diameter (Figure 3), it is clearly evident that lower cutting speed is necessary for all drill diameter values in the range 16 -28 mm in order to minimize the burr height.

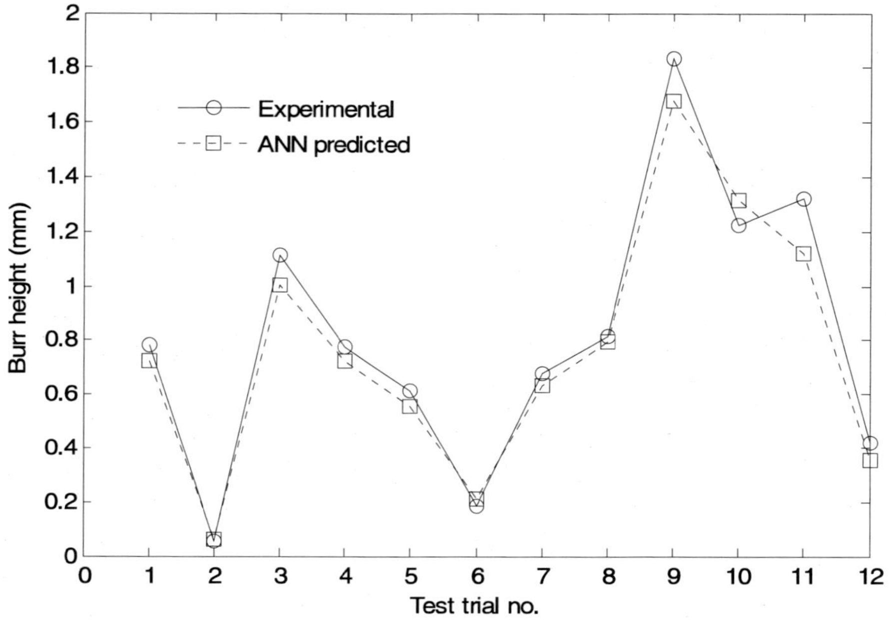

Figure 1. Comparison of the experimental and the ANN predicted values of burr height for validation data set.

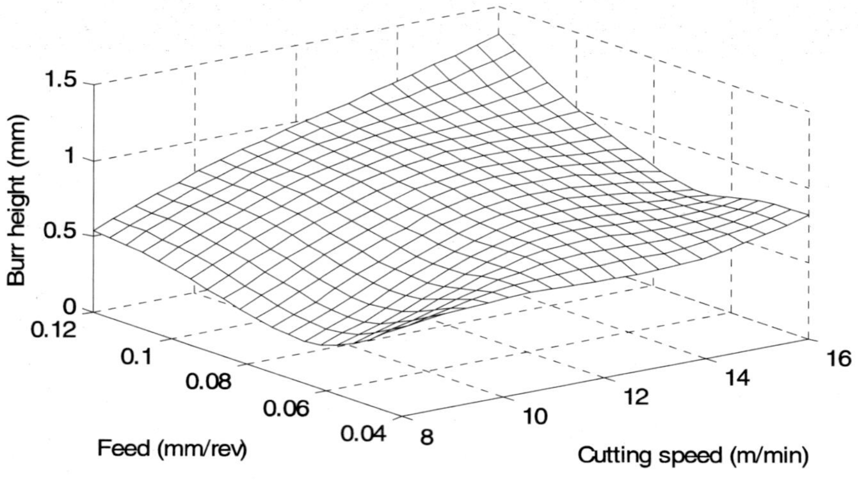

Figure 2. Interaction effect of cutting speed and feed on burr height.

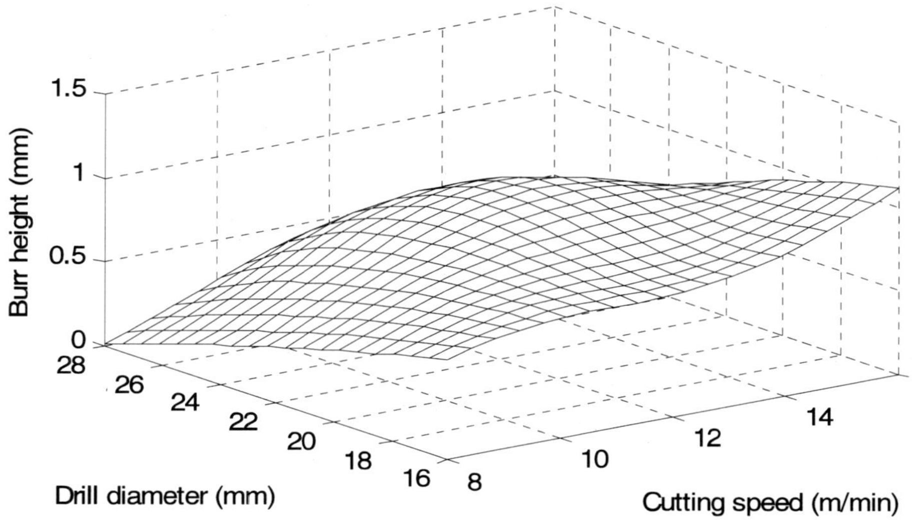

Figure 3. Interaction effect of cutting speed and drill diameter on burr height.

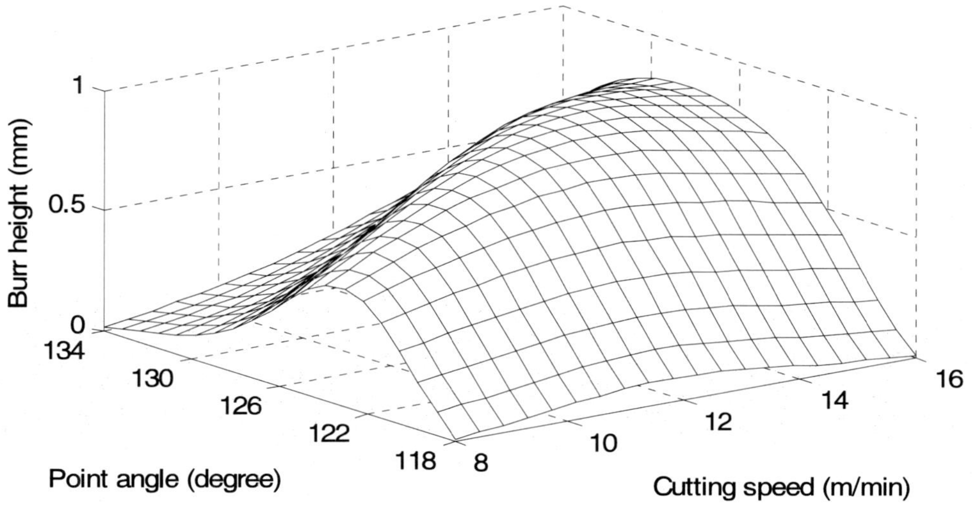

Figure 4. Interaction effect of cutting speed and point angle on burr height.

Figure 4 illustrates the interaction effects of cutting speed and point angle on burr height. It is seen from this figure that the burr height is less sensitive to cutting speed variations irrespective of the point angle. On the other hand, maximum burr height results when the point angle is in the range 122-130 degree for all values of cutting speed.

The interaction effects of cutting speed and lip clearance angle on burr height (Figure 5) suggests that the lower values of lip clearance angle and cutting speed are necessary for minimizing the burr height. In other words, the burr height increases either with the increase in cutting speed or with the increase in lip clearance angle.

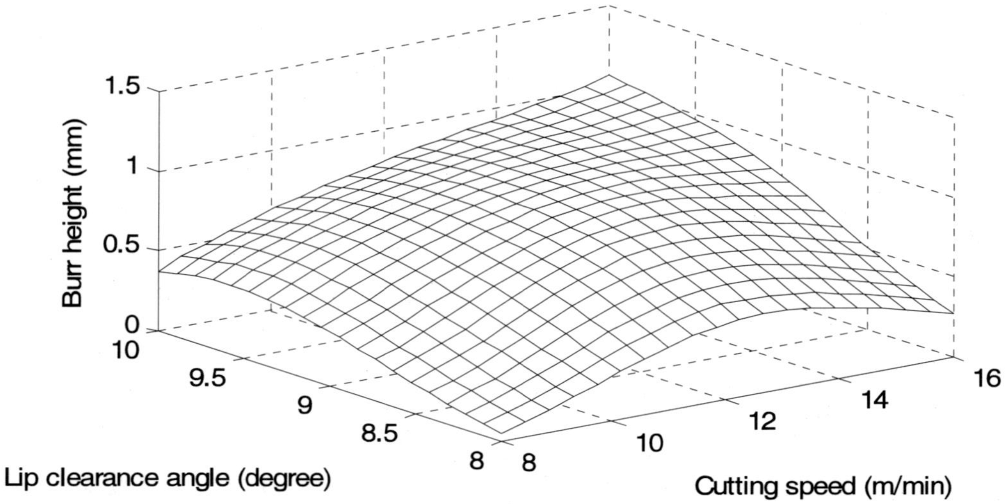

Figure 5. Interaction effect of cutting speed and lip clearance angle on burr height.

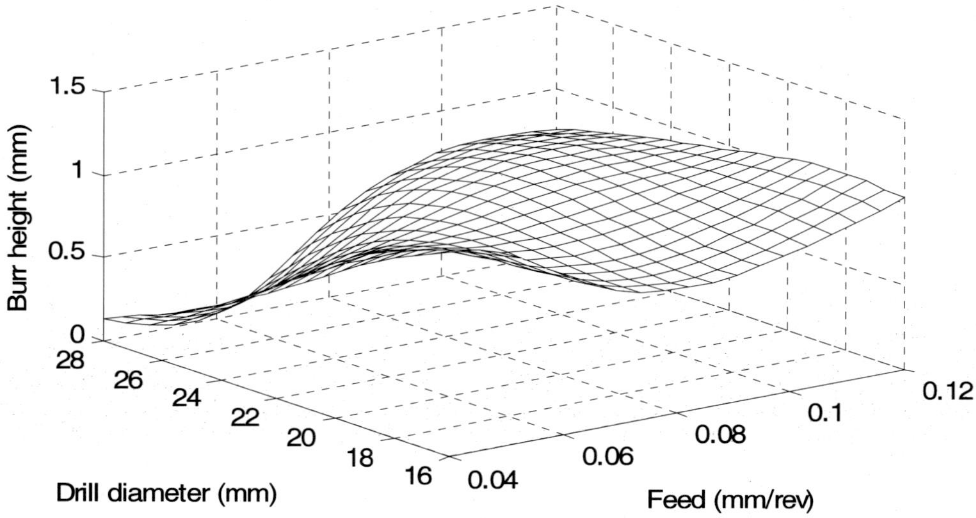

Figure 6. Interaction effect of feed and drill diameter on burr height.

As seen from Figure 6, the burr height is almost insensitive to feed variations for drill diameters up to 22 mm. However, beyond 22 mm drill diameter, lower feed value is beneficial for minimizing the burr height. It can be observed from Figure 7 that for any selected feed value, lower point angle is necessary to minimize the burr height. Moreover, for a specified feed, initially the burr height sharply increases with the point angle up to 126 degree. The interaction behavior of burr height with feed and lip clearance angle (Figure 8) is almost same as that with lip clearance angle and cutting speed (Figure 5).

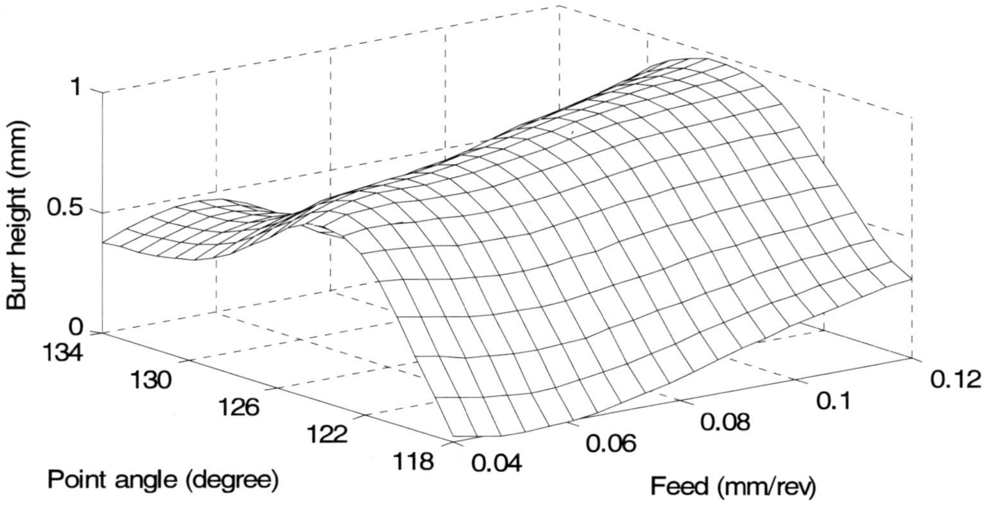

Figure 7. Interaction effect of feed and point angle on burr height.

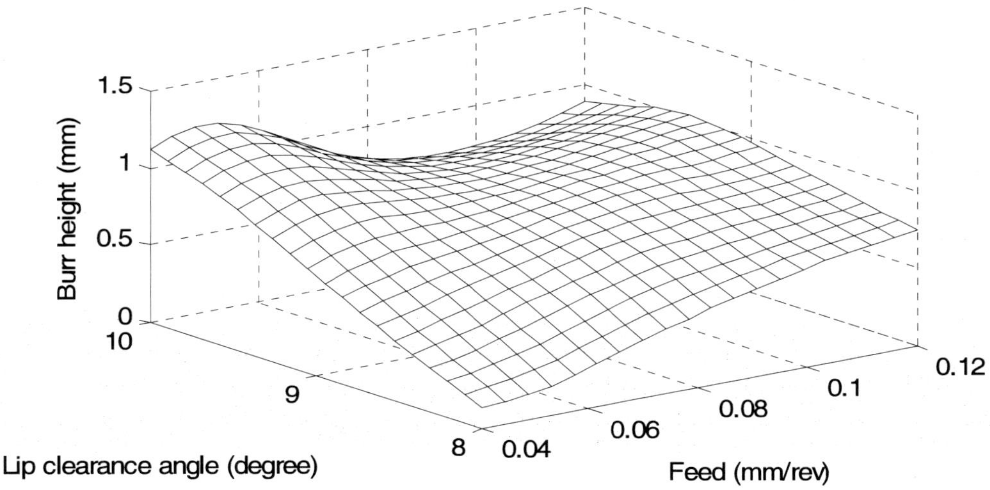

Figure 8. Interaction effect of feed and lip clearance angle on burr height.

It is clear from Figure 9 that, lower point angle is necessary to minimize the burr height for all drill diameters in the range 16-28 mm. Further, it is also observed that low lip clearance angle is essential as shown in Figure 10. Figure 11 exhibits the interaction effect of point angle and lip clearance angle. For any given lip clearance angle, the burr height shraply increases with point angle up to 126 degree and then shows decreasing tendency for further increase in point angle. Therefore, lower point angle is found to be suitable for minimizing the burr height.

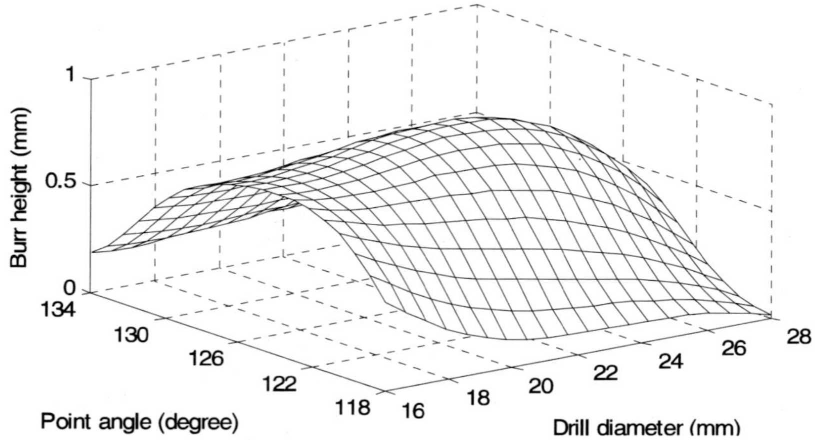

Figure 9. Interaction effect of drill diameter and point angle on burr height.

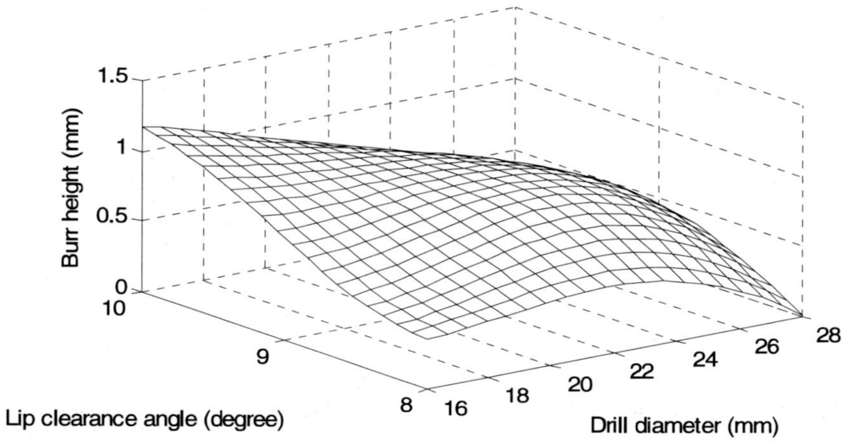

Figure 10. Interaction effect of drill diameter and lip clearance angle on burr height.

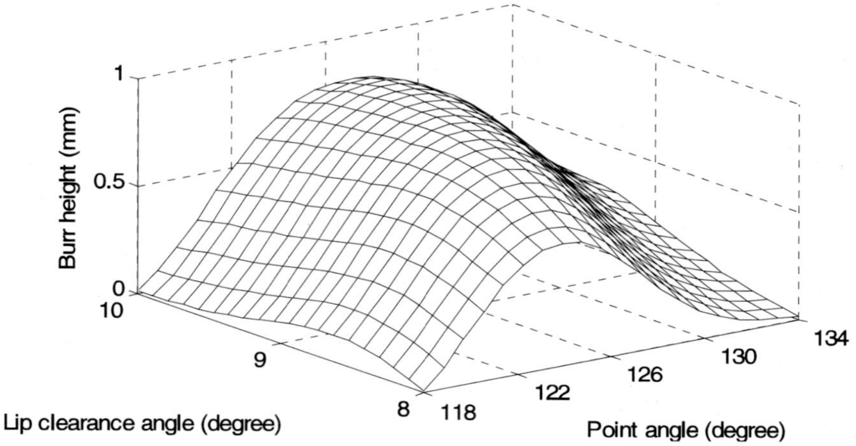

Figure 11. Interaction effect of point angle and lip clearance angle on burr height.

From the above discussion, it is clearly evident that the ANN model is very useful in analyzing the effects of drilling process parameters on burr height. With the developed model, it is possible to predict the burr height for a given combination of selected process parameters. Moreover, the developed ANN model is very useful for the selection of appropriate control parameters, which result in minimum burr height. The principal advantage of ANN modeling is that it has the ability to capture any degree of non-linearity that exists between the process parameters and the response with good generalization ability.

CONCLUSION

The multilayer feed forward artificial neural network (ANN) architecture; trained using error back propagation algorithm (EBPTA) was developed to analyze the effects of process parameters on burr height during drilling of AISI 316 stainless material. The input-output patterns required for ANN training were obtained through drilling experiments planned as per Box-Behnken design (BBD). A good agreement was observed between the ANN predicted and the experimental values of burr height. The simulated results demonstrated the effectiveness of the ability of ANN model in realizing a non-linear relationship between the burr height and the identified drilling process parameters. The following conclusions are drawn from the analysis:

- There exists a highly non-linear relationship between the burr height and the selected drilling process parameters.
- The burr height increases with increase in feed and is highly sensitive at higher values of cutting speed and feed. The minimum burr height results with lower values of cutting speed and feed combination.
- For minimizing the burr height, lower values of cutting speed are necessary for all the selected drill diameters. Further, lower point angle is necessary for minimizing the burr height for any given feed and for all drill diameters in the range 16-28 mm.
- The burr height increases either with increase in cutting speed or with increase in lip clearance angle. The lower feed value is beneficial for minimizing the burr height beyond 22 mm drill diameter.
- The maximum burr height exists when the point angle is in the range 122-130 degree for all values of cutting speed.

REFERENCES

[1] Stein, J. M. (1997), "*The Burrs from Drilling: An Introduction to Drilling Burr Technology*", Burr Technology Information Series ™, USA.

[2] Gaitonde, V. N., Karnik, S. R., Achyutha, B. T., Siddeswarappa, B. and Davim, J. P. (2009), "Predicting burr size in drilling of AISI 316L stainless steel using response surface analysis", *International Journal of Materials and Product Technology*, 35(1/2), 228-245.

[3] Schalkoff, R. B. (1997), "*Artificial Neural Networks*", McGraw-Hill International Edition.
[4] Montgomery, D. C. (2004), "*Design and Analysis of Experiments*", 5th edition, John Wiley and Sons, New York.
[5] Math Works Incorporation (2005), "*MATLAB User Manual Version 7.1 R14*", Math Works Incorporation, Natick, MA.

In: Artificial Intelligence in Manufacturing Research
Editor: J. Paulo Davim
ISBN 978-1-62618-557-9

Chapter 5

Artificial Neural Network Modeling of Surface Quality Characteristics in Abrasive Water Jet Machining of TRIP Steel Sheet

N. M. Vaxevanidis*[*,1], *A. Markopoulos*[1] and *G. Petropoulos*[2]

[1]Department of Mechanical Engineering Educators,
School of Pedagogical and Technological Education (ASPETE),
N. Heraklion, Athens, Greece
[2]Department of Mechanical and Industrial Engineering,
University of Thessaly, Pedion Areos, Volos, Greece

Abstract

In this chapter the application of Artificial Neural Networks (ANNs) for the modeling of surface quality characteristics (kerf geometry and surface roughness) in Abrasive Water Jet Machining (AWJM) of transformation induced plasticity (TRIP) steel sheet is discussed. For the development of the models the neural network toolbox of Matlab is used and all networks are feed-forward models trained with the Levenberg-Marquardt algorithm. Models constructed possess three inputs, namely nozzle diameter, stand-off distance and travel speed, one hidden layer with five neurons and two outputs (kerf and R_a). Training of the models was performed with data from an extensive series of statistically designed experiments concerning AWJM of two TRIP type steel sheets. The reported results indicate that the proposed ANN model can satisfactorily predict the surface roughness and the mean kerf in AWJM; moreover, it can be considered as valuable tools for process planning in workshop.

* Corresponding author, E-mail: vaxev@aspete.gr

INTRODUCTION

In abrasive water jet machining (AWJM), high velocity water containing abrasive particles is used to cut different materials ranging from soft, ductile to hard and brittle materials. AWJM technology nowadays is considered to be one of the most developing advanced non-traditional methods used in industry for material processing with the distinct advantages of no thermal distortion, high machining versatility, high flexibility and small cutting forces. Because of these capabilities, it can be applied for machining materials with higher performance and be more cost-effective than traditional and some non-traditional machining processes. AWJM is widely used in the processing of materials such as titanium, steel, brass, aluminium, stone, Inconel, any kind of glass and composites [1, 2].

The intensity and the efficiency of the machining process depend on several AWJM process parameters. They are classified as hydraulic, abrasive, work material and cutting parameters [3, 4]. Surface roughness, which is used to determine and to evaluate the surface quality of a product, together with the kerf geometry formed in a through cut [5], are the two major quality attributes of an AWJMed product which were evaluated during present research. For the estimation and/or the prediction of surface roughness in machining processes various methodologies have been adopted; for an overview see [6]. These different approaches are based on machining theory and/or process modeling, experimental investigation, design of experiments [7] and intelligent techniques. The present trend of research in the field of manufacturing favours intelligent techniques, probably due to the enhanced computing power. They consider the machining conditions and the information stored in the experimental data to develop the models. The various intelligent techniques attempted by researchers are (artificial) neural network (ANN), fuzzy logic, neuro-fuzzy approach and genetic algorithms [8, 9]. Following the succesful implementation of ANNs for R_a prediction due to electro-discharge machining (EDM); see Refs [9, 10], in this chapter the application of Artificial Neural network (ANN) for the modeling of surface quality characteristics (kerf geometry and surface roughness) in Abrasive Water Jet Machining (AWJM) of TRIP steel sheet is presented.

In brief, the structure of this chapter reads as follows: In the subsequent two sections the main features of AWJM and ANNs are, respectively, summarized. Then, the experimental features (materials, measuring techniques and the structure of ANNs selected) are outlined. Next, the application and the suitability of the proposed modeling are presented and discussed (Results and Discussion) and the representative literature follows.

ABRASIVE WATER JET MACHINING

The so-called non conventional (or non-traditional or advanced) machining processes have been emerged from the need in modern manufacturing practice for a more frequent use of harder, tougher or stronger workpiece materials, which are much more difficult to machine with traditional methods. Reference is made to all kinds of high strength thermal resistant alloys, to various kinds of carbides, fiber-reinforced composite materials, ceramic materials, various modern composite tool materials etc [1, 11].

These advanced machining processes become still more important when one considers precision and ultra-precision machining. In a number of processes, material is removed even in the form of atoms or molecules individually or in groups. These processes are based on the direct application of energy for material removal by mechanical erosion, thermal erosion or electro-chemical/chemical dissolution [12].

In general, advanced non conventional machining processes are classified into four main groups, namely:

- Mechanical machining processes
- Thermal machining processes
- Electro-chemical and chemical machining processes
- Hybrid machining processes

Among these processes abrasive water jet machining (AWJM) is regarded as the fastest growing major machine tool process in the world [13]. AWJM is a technique for cutting or separating materials by means of a high-velocity slurry jet, formed as a result of injecting abrasive particles to a waterjet ejected by an orifice; see Figure 1. Note, that pure water jet cutting was introduced around 1970 and the abrasive water jet cutting in 1983.

The main advantages of the AWJM process are being able to cut versatile geometries and its ability to cut both ductile materials like aluminum, brass, steel and titanium and brittle materials like glass, stone and ceramics without any influence on their microstructure. The water pressure, combined with the water flow rate defines the ability of the AWJM, since both factors define the maximum available power of the abrasive-water mixture [11, 14].

Note that the physical principles and the phenomena involved, the process parameters and the capabilities and applications of AWJM as well as recent advances and the limitations of the process can be found in a number of excellent reference works, see for example [1, 4, 15]; therefore only some main features are summarized below.

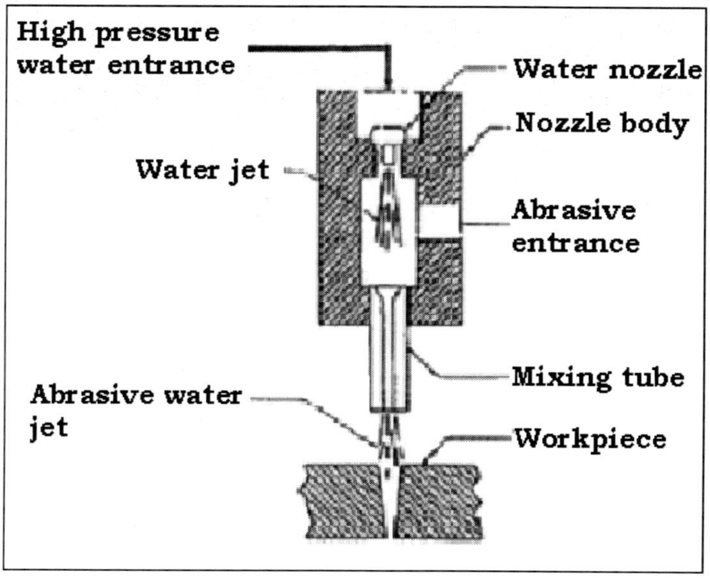

Figure 1. Schematic illustration of AWJM.

The abrasive water jet-cutting process is characterized by a large number of operational parameters which determine the efficiency, economy and quality of the entire process. In general, the parameters in AWJM can be divided into four categories [4]:

1. Hydraulic parameters
   - Pump pressure (p)
   - Water-orifice diameter
   - Water flow rate
2. Mixing and acceleration parameters
   - Focus diameter
   - Focus length
3. Cutting parameters
   - Traverse rate
   - Number of passes (n_p)
   - Stand off distance (x)
   - Impact angle
4. Abrasive parameters
   - Abrasive mass flow-rate
   - Abrasive particle diameter
   - Abrasive particle size distribution
   - Abrasive particle shape
   - Abrasive particle hardness

Material removal in AWJM of metals is comprised of microchip formation, plowing and rubbing, all of which occur through shear deformation. The predominance of any one of these three mechanisms is depended on the abrasive attack angle and the mechanical properties of the target material; for an overview see [2, 15]. At low cutting depths, abrasives impinge on the target material at shallow attack angles, hence promoting "cutting wear". The impingement angle was believed to become more obtuse with larger cutting depths, thereby inducing a change of removal mechanisms to "deformation wear". Material removal within the deformation wear region occurs through a cyclic cutting action and commonly results in a wavy surface texture which serves as the division between the cutting wear and deformation wear regions.

The pressure at which a water jet operates is about 400 MPa, which is sufficient to produce a jet velocity of 900 m/s. Such a high-velocity jet is able to cut materials such as ceramics, composites, rocks, metals etc. [1]. The AWJM process can easily cut both electrically non-conductive and conductive, and difficult-to-machine materials. This process does not produce dust, thermal defects, and fire hazards. Recycling of water and abrasives is possible to some extent. It is a good process for shaping and cutting of composite materials, and creates almost no delamination [4, 12].

The kerf geometry of a through cut generated by abrasive waterjets may be described as in Figures 2(a) and 2(b). It is characterized by a small rounded corner at the top edge due to the plastic deformation of material caused by jet bombardment. As the kerf is wider at the top than at the bottom due to the decrease in water pressure, a taper is produced. In addition, the plastically deformed material rolls over at the bottom of the kerf forming burrs at the jet exit when cutting ductile materials.

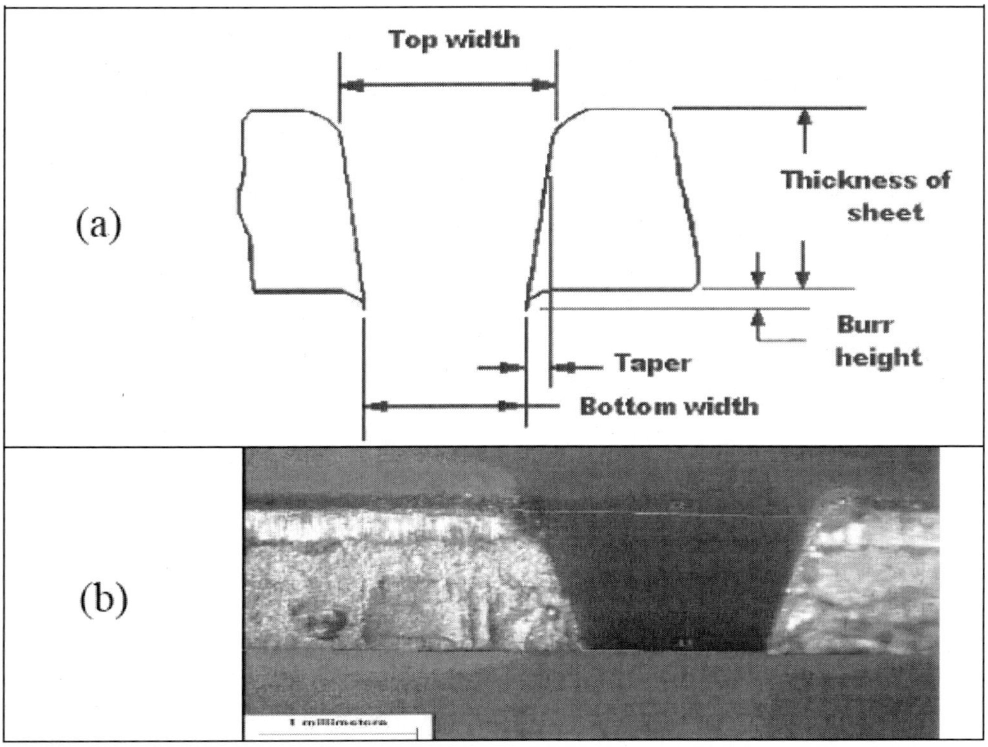

Figure 2. Geometry of a typical cut in AWJM; (a) a schematic representation and (b) macrograph of a cut sheet.

The primary interests in sheet steel processing are the kerf shape (kerf width and kerf taper) and kerf quality (cut surface roughness) as well as burrs which may be formed at the jet exit [16]. These characteristics were considered in the present study; see next section "experimental".

ARTIFICIAL NEURAL NETWORK

Artificial Neural Networks Overview

Two main and important features of neural networks are their architecture, i.e., the way that the network is structured, and the algorithm used for its training. After the appropriate training, the selected network has the ability to interconnect one value of output to given input. These two features of neural networks along with some techniques used for the improvement of their performance are briefly presented below. Note that the origin, the development and the mathematical details for implementing the ANNs can be found in a number of excellent reference works, see for example Refs [17-19]; therefore they are not discussed here.

Neural Network's Architecture

Artificial neural networks are mathematical representations of the human brain function. The "core" element of a neural network is the neuron. Neurons are connected to each other with a set of links, called synapses and each synapse is described by a synaptic weight. Neurons are placed in layers and each layer's neurons operate in parallel. The first layer is the input layer. The activity of input units represents the non-processed information that entered the network; at that layer neurons do not perform any computations. The hidden layers follow the input layer and the activity of each hidden unit is determined from the activity of the input units and the weights at the connections of input and hidden units. A network can have many or none hidden layers and their role is to improve the network's performance. The existence of these layers at the network becomes more necessary as the number of input neurons grows. The last layer is the output layer. The behavior of output units depends upon the activity of the hidden units and the weights between hidden units and output units. The output of the layer is the output of the whole network; output layer neurons, in contrast to input layers, perform calculations.

There are two types of neural networks: the feed-forward and the recurrent ones. Feed-forward neural networks allow the signals to travel in only one direction: from input to output, i.e. the output signal of a neuron is the input of the neurons of the following layer and never the opposite. The inputs of the first layer are considered the input signals of the whole network and the output of the network is the output signals of last layer's neurons. On the contrary, recurrent networks include feedback loops allowing signals to travel forward and/or backward [18]. Feed-forward neural networks are characterized by simple structure and easy mathematical description [19]; therefore they were selected for the modeling of surface roughness in the present chapter.

In general, there is not a standard algorithm for calculating the proper number of hidden layers and neurons. For relatively simple systems, as the present case, a trial-and-error approach is usually applied in order to determine which architecture is optimal for a problem. Networks that have more than one hidden layers have the ability to perform more complicated calculations. However, for most applications, one hidden layer is enough, while for more complicated applications the simulation usually takes place using two hidden layers [20]. The existence of more than necessary hidden layers complicates the network, resulting in a low speed of convergence during training and large error during operation. Therefore, the architecture of a neural network always depends upon the specific situation examined and must not be more complex than needed [10, 17].

Neural Network's Training

Once the number of layers and the number of units in each layer are selected the network's weights must be set in order to minimize the prediction error of the network; this is the role of the training algorithms. The historical cases that were gathered are used to automatically adjust the weights in order to minimize this error. The error of a particular configuration of the network can be determined by running all the training cases through the network and comparing the actual output generated with the desired or target outputs. The differences are combined together by an error function resulting the network's error. Usually the mean square error (MSE) of the network's response to a vector p, is calculated, according to the equation:

$$E_p = \frac{1}{2}\sum_{i=1}^{l}\left(d_{p,i} - o_{p,i}\right)^2$$

In the preceding equation $o_{p,i}$ are the values of the output vector which occur for the input vector p and $d_{p,j}$ the values of the desirable response corresponding to p. The procedure is repeated until MSE becomes zero. Each time that the program passes through all pairs of training vectors an epoch is completed; training usually ends after reaching a great number of epochs.

One of the frequently used training algorithms is the back-propagation (BP) algorithm. It is usually applied in feed-forward networks with one or more hidden layers [10, 21, 22]. The input values vectors and the corresponding desirable output values vectors are used for the training of the network until a function is approached which relates the input vectors with the particular output vectors. When the value of the mean square error is calculated, it is propagated to the back in order to minimize the error with the appropriate modification of the weights.

Another important parameter of the neural network models is their ability to generalize, i.e. the ability of neural networks to provide logic responses for input values that were not included in the training. Correctly trained back-propagation networks are able to perform generalization; this ability provides the opportunity of training the network using a representative set of input - desirable output values pairs.

Improvement Techniques

When an algorithm is applied to the network random values are given to the weight factors. The convergence speed and the reliability of the network depend upon the initial values of weights; thus different results may be observed during the application of the same algorithm to the network. There are only a few elements that can guide the user for the selection of the proper values. A wrong choice may result to small convergence speed or even to network's paralysis, where training stops. Furthermore, due to the fact that the algorithm searches for the minimum error, the network may be stabilized at a local minimum instead of the total minimum. As a result, most of the times, incorrect response values of the network are produced. To overcome these problems variations of the most used algorithms have been proposed; for further information on this topic Refs. [18, 19] may be consulted. Worth mentioning, also, that a very common and simple technique used for overcoming problems of this type is the repetition of the algorithm many times and the use of different initial values of the weight factors.

One of the problems that occur during the training of neural networks is over-fitting which undermines their generalization ability. The error appears to be very small at the set of the training vectors, however, when new data are imported to the network the error may become extremely large. This phenomenon is attributed to fact that the network memorized the training examples and did not learn to generalize under the new situations. The generalization ability of a network is assured when the number of training data is quite greater than the number of network's parameters. However, when the network is large the relations between input and output become rather complicated. Hence, a network should not be larger than needed to solve the given problem. Note, also, that two improvement techniques were applied during modeling, namely, normalization of the used data and the early stopping technique. Both these techniques and their application to the particular problem are briefly discussed in section: Experimental/ANN modeling.

Note that in the area of machining, neural networks have been used for the prediction of cutting forces, surface roughness, dimensional deviation, tool wear and tool life; see for an overview [23]. To the authors' knowledge, applications of ANNs to AWJM are quite limited; see Refs. [22, 24, 25].

EXPERIMENTAL

Abrasive Water Jet Machining

Machining was performed on a SIELMAN HELLENIC HYDROJET industrial AWJM system. In each specimen three slots of 3 cm in length were cut on both sides; each slot corresponds to different machining conditions. After processing, each specimen was separated in order to allow roughness measurements to be performed on the machined surface. The specimens were processed with three different diameters of the nozzle (nozzle diameter), three different distance values between the nozzle and the sheet steel (stand off distance) and three different cutting speeds (travel speed). Concisely, the design variables for AWJM modeling are given in the following table (Table 1). Worth mentioning that the high value of stand off distance is not usually applicable but was considered to establish a wide range of variation.

Table 1. Design variables of the machining process.

Nozzle diameter (mm)	Stand-off distance (mm)	Travel speed (mm/min)
0.95	20	200
1.2	64	300
1.5	96	400

Materials

TRIP (Transformation Induced Plasticity) multi-phase steels belong to a new generation of steel grades exhibiting an enhanced combination of strength and ductility, with extensive applications in automotive and aerospace industry.

The TRIP steels tested are designated TRIP 800 HR-FH and TRIP 700 CR-FH. Their chemical composition is summarized as follows.

- TRIP 800 HR-FH: 0.22 % C, 0.04 % Si, 1.66 % Mn, 1.49 % Al, 0.012 % P, 0.001 % S, 0.027% Cr, 0.003 %Mo, 0.02 %Ni and 0.027 %Cu.
- TRIP 700 CR-FH: 0.21 % C, 1.53 % Si, 1.78 % Mn, 0.04 % Al, 0.019 % P, 0.003 % S, 0.026% Cr, 0.001 % Mo and 0.027 % Cu.

Specimens of both materials are of square form (10x10 cm^2) but differ in thickness, hardness and processing method; see Table 2.

Table 2. Properties of materials tested

Material	TRIP 800 HR-FH	TRIP 700 CR-FH
Thickness	1.25 mm	0.9 mm
Hardness	20 HRC	23 HRC
Processing method	Hot rolling	Cold rolling

Measuring Techniques

The roughness measurements were carried out on twenty-seven specimen surfaces using Surtronic 3+ stylus profilometer supported by the commercial software Talyprof®. The cut-off selected was 0.8 mm and the measurements were undertaken in the direction of the cut. The parameter values appear as averages of 5 measurements on each surface at the medium area of the cut. Although, there is serious surface inhomogeneity between the entrance and the exit areas of the beam, these values are considered as representative for the quality control of the AWJMachined surfaces. In Figure 3 a typical profile is shown, indicative of the striated pattern generated by AWJM; see also [26].

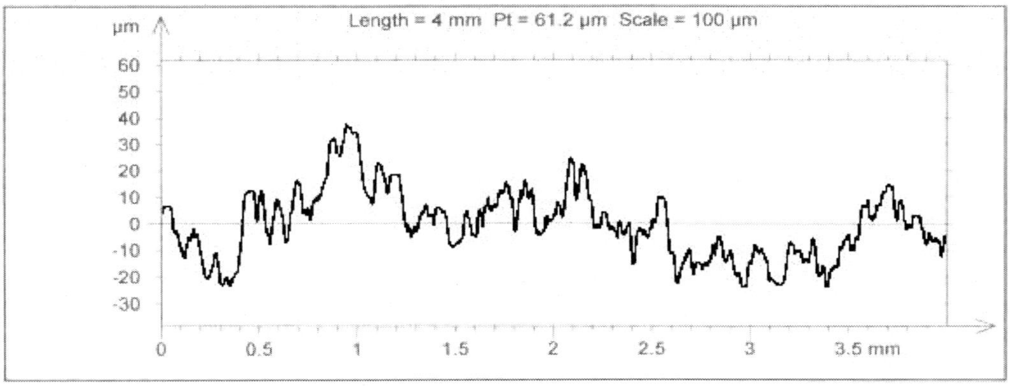

Figure 3. Typical roughness profile of AWJMed surface.

As it is illustrated in Figure 2(b) the kerf is of tapered form and to evaluate this characteristic, the semi-sum of the upper area width and the lower area width was employed.

ANN Modeling

Two different neural networks models were constructed for each steel type, namely TRIP 800 and TRIP 700. Each model can predict "mean kerf diameter" (Kerf) and "mean surface roughness" (R_a). These models were constructed using Matlab® with the neural network toolbox (version R2008a) [27].

All networks are feed-forward models trained with the Levenberg-Marquardt algorithm. They have three inputs, namely nozzle diameter, stand-off distance and travel speed, one hidden layer with five neurons and two outputs (kerf and R_a). The transfer function in the hidden layer is the hyperbolic tangent sigmoid function. The architecture of the selected (optimized) network is presented in Figure 4.

For each model the early stopping technique is applied. In this method the existing data are divided into three subsets. The first subset consists of the training vectors, which are used to calculate the gradient and to form the weight factors and the bias. The second subset is the validation group. The error in this group is observed during training and like training group normally decreases during the initial phase of training. However, when the network begins to adjust the data more than needed, the error in that group raises and when this increase is continued for a certain number of repetitions, training stops. The third subset is the test group and its error is not used during training. It is used to compare the different models and algorithms. The number of data as well as which data are included in every group is randomly selected by the program.

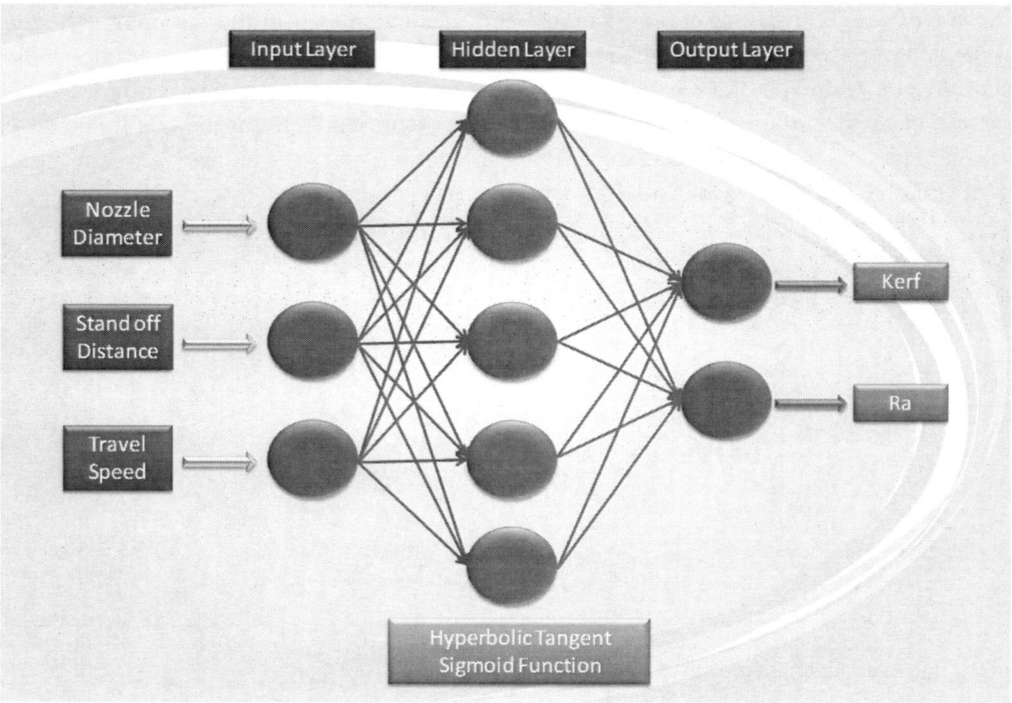

Figure 4. Optimized neural network architecture.

Furthermore, Principal Component Analysis is used in the research outlined in the present chapter. In some situations, the dimension of the input vector is large, but the components of the vectors are highly correlated (redundant). It is useful in this situation to reduce the dimension of the input vectors. An effective procedure for performing this operation is principal component analysis. This technique has three effects: it orthogonalizes the components of the input vectors (so that they are uncorrelated with each other), it orders the resulting orthogonal components (principal components) so that those with the largest variation come first, and it eliminates those components that contribute the least to the variation in the data set. In the present analysis, firstly the input vectors are normalized, so that they have zero mean and unity variance. Then the principal components that contribute less than 1% to the total variation in the data set are eliminated.

RESULTS AND DISCUSSION

Based on the design variables for AWJM modeling presented in Table 1, the process parameters and the corresponding measured values for R_a and kerf are tabulated in Tables 3 and 4 for TRIP 800 and TRIP 700 steel grades respectively. In the same Tables the results from the proposed ANN as well as the deviation ("Dev.") between measured and calculated by ANN values are presented.

Table 3. Results for TRIP 800 steel grade

Nozzle diameter (mm)	Stand off distance (mm)	Travel speed (mm/min)	Kerf (mm)	Kerf (by ANN) (mm)	Dev. (%)	R_a (μm)	R_a by ANN (μm)	Dev. (%)
0.95	20	200	0.915834	0.9337	1.95079	5	5.1112	2.224
0.95	20	300	0.872053	0.9506	9.007136	5.3	5.4966	3.709434
0.95	20	400	0.857991	1.0182	18.67257	5.9	5.7055	3.29661
0.95	64	200	1.113319	0.8841	20.5888	6	6.0626	1.043333
0.95	64	300	1.0768	0.9828	8.729569	5.7	5.7161	0.282456
0.95	64	400	0.96584	1.0425	7.937132	6.7	6.0615	9.529851
0.95	96	200	1.164883	1.1374	2.359293	6.1	6.0564	0.714754
0.95	96	300	0.952434	1.0155	6.621561	6.3	6.3097	0.153968
0.95	96	400	0.924122	0.9993	8.135073	6.6	6.5694	0.463636
1.2	20	200	1.18478	1.1435	3.484191	6.1	6.0777	0.365574
1.2	20	300	1.175643	1.1385	3.159377	6.3	6.2883	0.185714
1.2	20	400	1.048306	1.1523	9.920195	6.2	6.1874	0.203226
1.2	64	200	1.285587	1.2588	2.08364	6.8	6.6491	2.219118
1.2	64	300	1.322628	1.2069	8.749853	6.7	6.6936	0.095522
1.2	64	400	1.150741	1.1895	3.368178	6.6	6.5893	0.162121
1.2	96	200	1.447989	1.5168	4.752177	6.85	6.902	0.759124
1.2	96	300	1.434771	1.4156	1.336171	7.05	6.9841	0.934752

Table 3. (Continued)

Nozzle diameter (mm)	Stand off distance (mm)	Travel speed (mm/min)	Kerf (mm)	Kerf (by ANN) (mm)	Dev. (%)	R_a (μm)	R_a by ANN (μm)	Dev. (%)
1.2	96	400	1.26562	1.2695	0.306569	7	7.0369	0.527143
1.5	20	200	1.353131	1.2437	8.087244	6.1	6.1126	0.206557
1.5	20	300	1.224977	1.1939	2.536946	5.7	5.8553	2.724561
1.5	20	400	1.384728	1.1655	15.83185	5.75	6.3982	11.27304
1.5	64	200	1.501308	1.5067	0.359153	6.7	6.7411	0.613433
1.5	64	300	1.34832	1.421	5.390412	6.85	6.8475	0.036496
1.5	64	400	1.616713	1.3643	15.61273	6.75	6.5312	3.241481
1.5	96	200	1.58843	1.5798	0.543304	7.05	6.9039	2.07234
1.5	96	300	1.560049	1.5604	0.022499	6.8	6.9268	1.864706
1.5	96	400	1.516709	1.5101	0.435746	6.9	6.9479	0.694203

Table 4. Results for TRIP 700 steel grade

Nozzle diameter (mm)	Stand off distance (mm)	Travel speed (mm/min)	Kerf (mm)	Kerf (by ANN) (mm)	Dev. (%)	R_a (μm)	R_a by ANN (μm)	Dev. (%)
0.95	20	200	0.9780385	0.9994	2.184116	4.5	4.6297	2.882222
0.95	20	300	1.0265984	0.9859	3.964393	4.1	4.794	16.92683
0.95	20	400	0.9612056	1.0554	9.79961	5.2	5.692	9.461538
0.95	64	200	1.11525	1.1088	0.578346	6.25	5.9997	4.0048
0.95	64	300	1.1554065	1.1524	0.260211	6.15	6.0786	1.160976
0.95	64	400	1.088564	1.0386	4.5899	5.95	6.0815	2.210084
0.95	96	200	1.163348	1.2351	6.167716	6.45	6.6784	3.541085
0.95	96	300	1.1400259	1.2733	11.69044	6.5	6.6293	1.989231
0.95	96	400	1.0820467	1.0371	4.15386	7.1	7.0151	1.195775
1.2	20	200	1.350591	1.3847	2.525487	6.2	6.1749	0.404839
1.2	20	300	1.329437	1.4459	8.760325	6.1	6.1591	0.968852
1.2	20	400	1.3352925	1.3235	0.88314	6.3	6.2844	0.247619
1.2	64	200	1.4216195	1.4469	1.778289	6.45	6.678	3.534884
1.2	64	300	1.4231235	1.5675	10.14504	7.3	7.6855	5.280822
1.2	64	400	1.3685895	1.451	6.021565	8.3	7.7036	7.185542
1.2	96	200	1.610029	1.5636	2.883737	7.9	7.8212	0.997468
1.2	96	300	1.5572765	1.604	3.000334	8.8	7.9298	9.888636
1.2	96	400	1.4473185	1.4209	1.825341	8.8	8.4547	3.923864
1.5	20	200	1.651102	1.6122	2.356123	7	6.9934	0.094286
1.5	20	300	1.464466	1.7267	17.90646	7	7.2331	3.33
1.5	20	400	1.739387	1.6887	2.914073	7	7.1597	2.281429
1.5	64	200	1.8041005	1.7254	4.362312	7.35	7.2777	0.983673

Table 4. (Continued)

Nozzle diameter (mm)	Stand off distance (mm)	Travel speed (mm/min)	Kerf (mm)	Kerf (by ANN) (mm)	Dev. (%)	Ra (μm)	Ra by ANN (μm)	Dev. (%)
1.5	64	300	1.8291155	1.7522	4.205065	8.1	7.8708	2.82963
1.5	64	400	1.791895	1.7174	4.157331	8.65	8.6593	0.107514
1.5	96	200	1.8016555	1.7028	5.486926	9.05	7.8651	13.09282
1.5	96	300	1.710008	1.7094	0.035555	8.2	8.3314	1.602439
1.5	96	400	1.730972	1.6332	5.648387	8.3	9.215	11.0241

Modeling of TRIP 800 Steel Grade

The MSE of training of the selected ANN was about 0.0153 and its training took 11 epochs to complete. The performance of the model is clearly indicated in Figure 5. The best validation performance is 0.015296 at epoch 5. From this figure is also evident that validation and testing group MSEs are higher than that of the training group, as expected; see also [10].

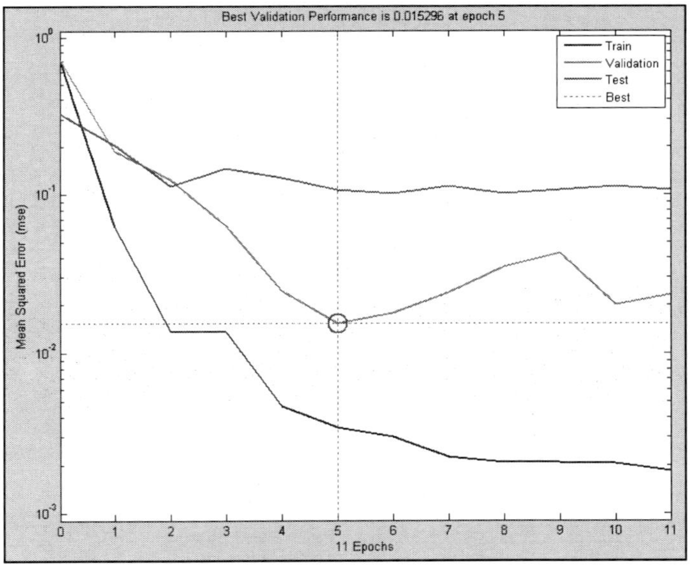

Figure 5. Results of the neural network training for AWJM of TRIP 800 grade.

For the evaluation of the generalization ability (quality) of the trained neural networks linear fits between the outputs of the model and the experimental data, for all the measurements, without discrimination to which group they belong, was performed. The graph of the linear fit is presented in Figure 6; note that "Target" represent the experimental results and "Output" the output values of the model. The best linear fit function is calculated as being: Output=0.81Target+0.22, while the correlation coefficient R is equal to 0.89883 for Kerf and Output=0.85Target+0.94, while the correlation coefficient R is equal to 0.93022 for R_a (with R=1 meaning that the best linear fit is achieved and the A=T curve match perfectly).

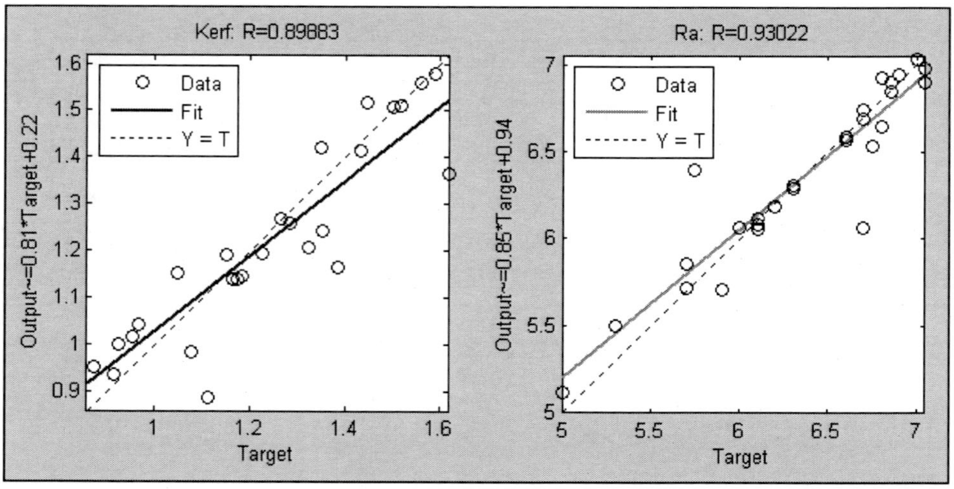

Figure 6. Correlation between experimental data and neural network output.

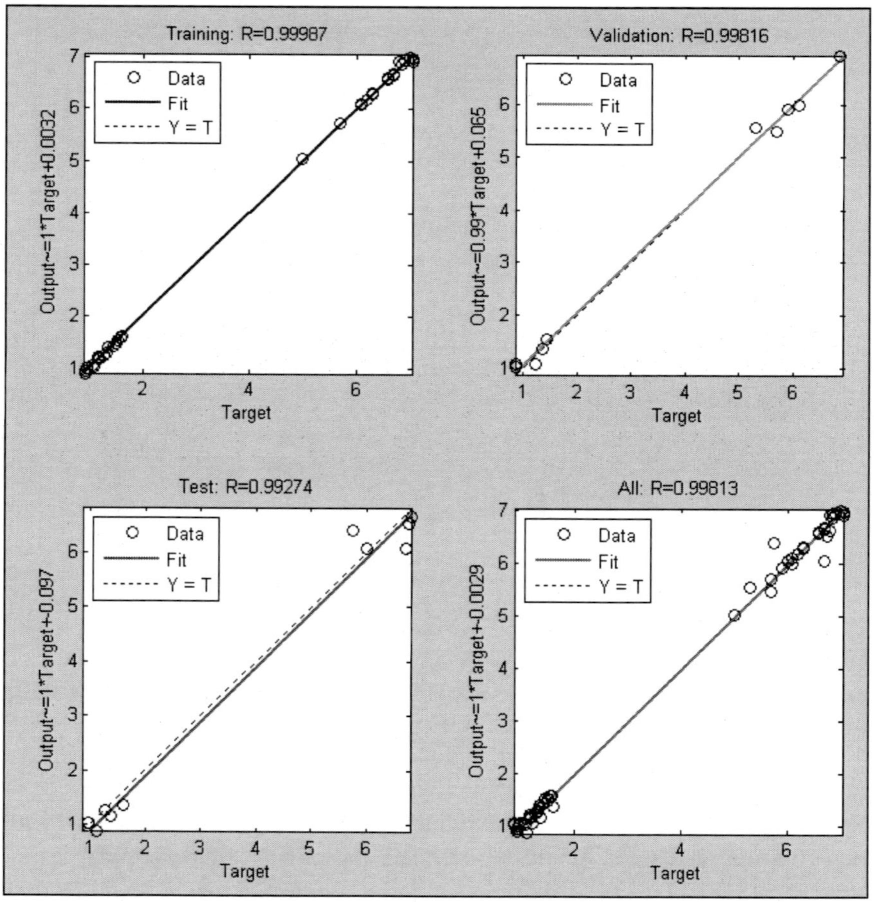

Figure 7. Regression analysis for the whole ANN model.

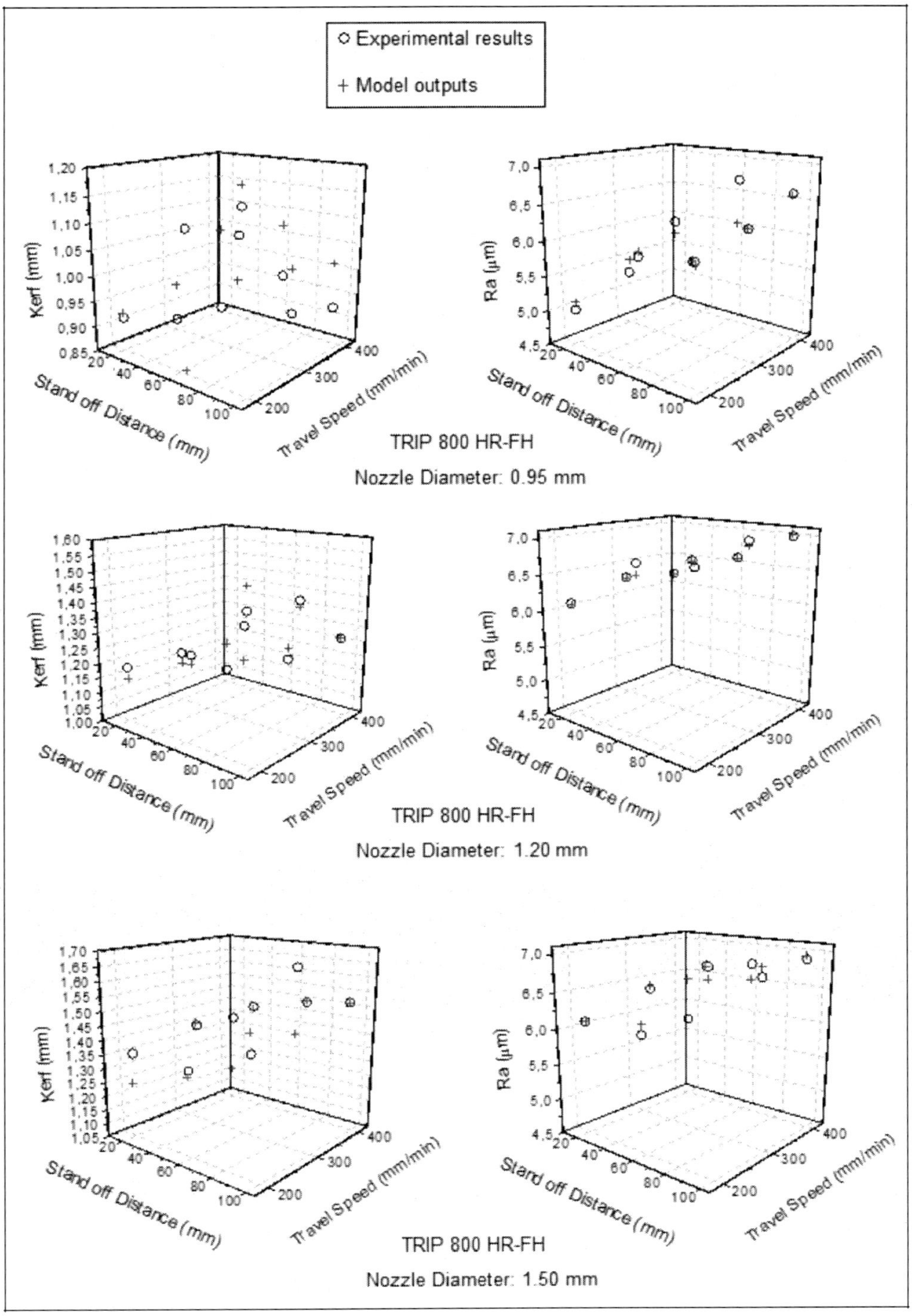

Figure 8. Experimental data and ANN results for R_a and kerf in AWJM of TRIP 800 steel grade.

A regression analysis was also performed for the whole model, see Figure 7. In the graphs of this figure, training, validation and test data group are depicted, first separately and then, all-in-one group.

Results in Figures 5-7 indicates that the proposed ANN can very satisfactorily predict the output data required. Furthermore, in Figure 8, 3-D plots of the centre-line average surface roughness (R_a) and the kerf versus the stand off distance and the travel speed are presented for all three nozzle diameters used. In these figures both the experimental data and the neural network outputs are presented. It is evident for all six plots that the experimental and the calculated values exhibit small discrepancies, indicating once more the reliability of the neural network constructed.

Modeling of TRIP 700 Steel Grade

Process parameters, measured values and the results from the proposed ANN applied in AWJM of TRIP 700 grade as well as the deviation ("Dev.") between measured and calculated by ANN values, are summarized in Table 4.

Results obtained from ANN modeling are qualitatively almost identical with the ones concerning TRIP 700 steel grade. The MSE of training of the selected ANN was about 0.271 and its training took 10 epochs to complete. The performance of the model is clearly indicated in the next Figure 9.

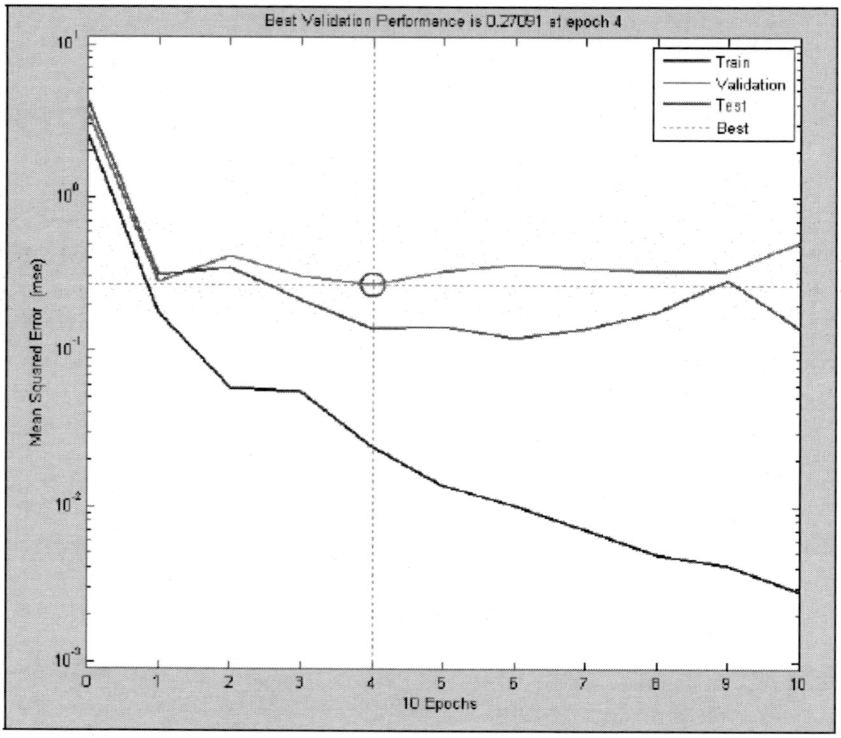

Figure 9. Results of the neural network training for AWJM of TRIP 700 grade.

The best validation performance is 0.27091 at epoch 4. From this figure is again evident that validation and testing group MSEs are higher than that of the training group, as expected. The graph of the linear fit is presented in Figure 10; note that "Target" represent the experimental results and "Output" the output values of the model. The best linear fit function is calculated as being: Output=0.88Target+0.17, while the correlation coefficient R is equal to 0.9533 for Kerf and Output=0.82Target+1.2, while the correlation coefficient R is equal to 0.94728 for R_a (with R=1 meaning that the best linear fit is achieved and the A=T curve match perfectly).

Results in Figures 9 and 10 indicate that the proposed ANN can very satisfactorily predict the output data required. Furthermore, in Figure 11, 3-D plots of the centre-line average surface roughness (R_a) and the kerf versus the stand off distance and the travel speed are presented for all three nozzle diameters used. In these figures both the experimental data and the neural network outputs are presented. It is evident for all six plots that the experimental and the calculated values exhibit small discrepancies; a behaviour which was already reported for TRIP 800 steel grade.

In general, ANN models for both TRIP 800 and TRIP 700 steel grades were proven to perform well for AWJM, giving reliable predictions and providing thus a possible way to avoid time- and money-consuming experiments.

Note also that modeling of AWJM by ANN could be combined with a Knowledge Based System (KBS) based on both rules and Object-Oriented structures such as Semantic Networks (SN) and Frames, which maintain, process and relate taxonomies of products and materials, their attributes and properties, processes' parameters, quality control specifications and manufacturability rules; see preliminary results on the topic reported in Ref. [28].

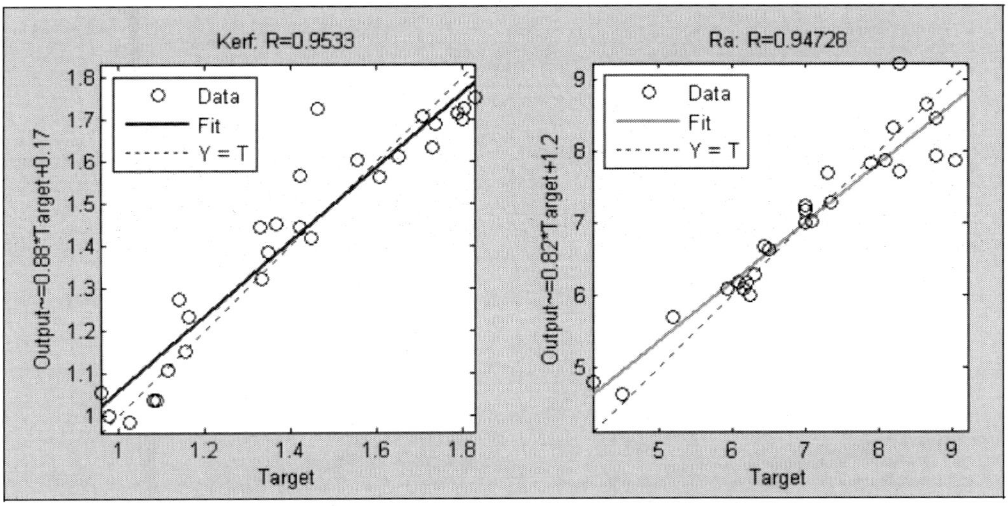

Figure 10. Correlation between experimental data and neural network output for TRIP 700 steel grade.

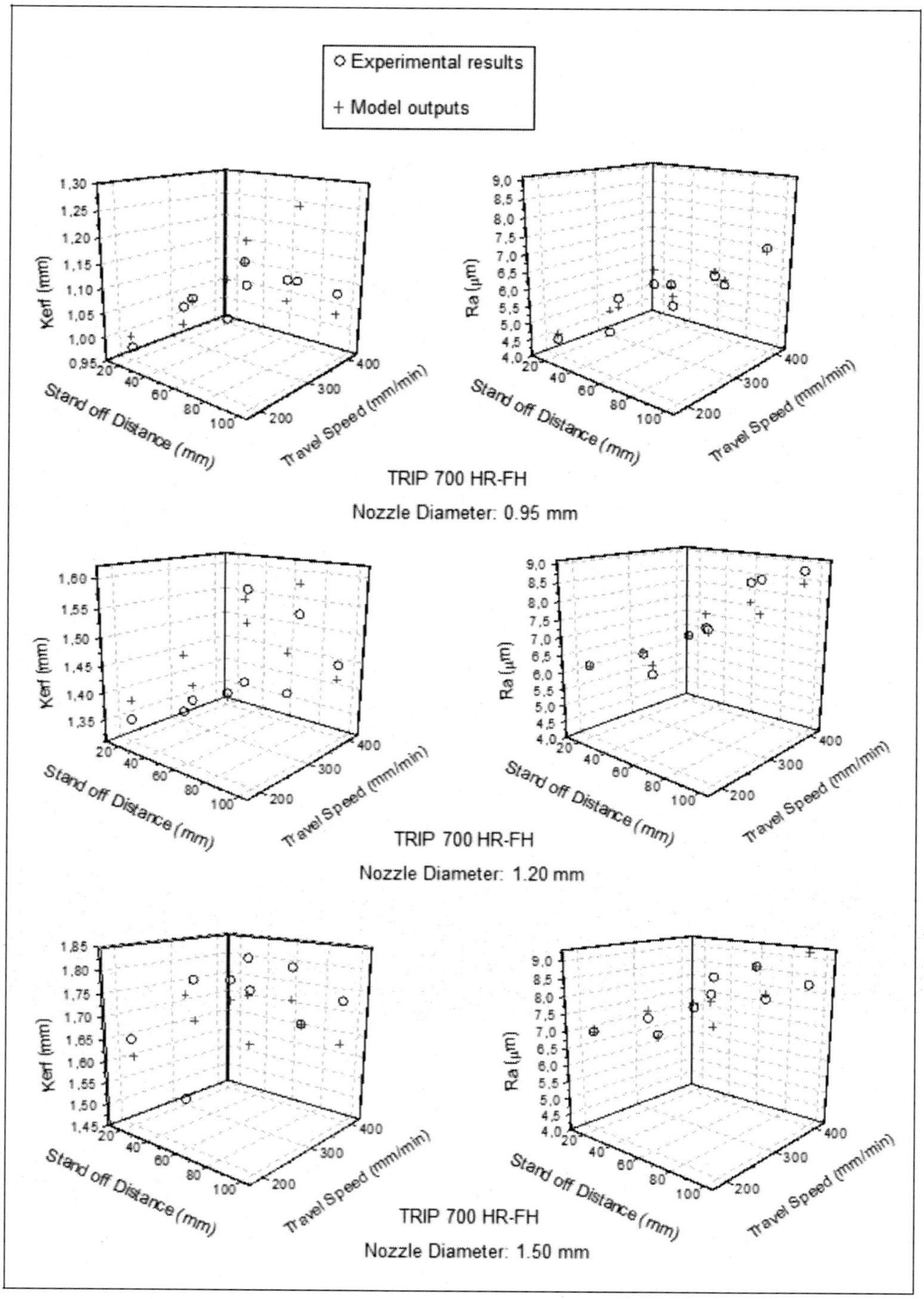

Figure 11. Experimental data and ANN results for R_a and kerf in AWJM of TRIP 700 steel grade.

Conclusion

Abrasive waterjet machining (AWJM) is a technique for cutting or separating materials by means of a high-velocity slurry jet, formed as a result of injecting abrasive particles to a waterjet ejected by a nozzle. The main advantages of the AWJM are being able to cut versatile geometries and its ability to cut both ductile materials like aluminium, brass, steel and titanium and brittle materials like glass, stone and ceramics without any influence on their microstructure. In sheet steel processing with AWJM the primary output characteristics, as far as the quality of the cut is concerned are the kerf shape and the cut surface roughness. The average kerf width and the centre-line average surface roughness (R_a) were considered in the present study. For the prediction of these features in AWJM two different ANNs were developed using the neural network toolbox of Matlab®. Models constructed possess three inputs, namely nozzle diameter, stand off distance and travel speed, one hidden layer with five neurons and two outputs (kerf and R_a). Training of the models was performed with data from an extensive series of statistically designed experiments concerning AWJM of two TRIP (transformation induced plasticity) type steel sheets.

The reported results indicate that the proposed ANNs models can satisfactorily predict the surface roughness and the mean kerf in AWJM; moreover, they can be considered as valuable tools for the process planning and moreover, provide a possible way to avoid time- and money-consuming experiments.

References

[1] Jain, VK. Advanced (Non-traditional) Machining Processes (ch. 11). In Davim, JP, editor. *Machining - Fundamentals and Recent Advances*. Springer; 2008; 299-327.

[2] Arola, D; Ramulu, M. Material removal in abrasive waterjet machining of metals Surface integrity and texture. *Wear*, 1997 210, 50-58.

[3] Hashish, M. Optimization factors in abrasive waterjet machining. *Journal of Engineering for. Industry (Tranactions of the. ASME)* 1991 113, 29–37.

[4] Momber AW; Kovacevic, R. *Principles of Abrasive Water Jet Machining*, Springer, 1997.

[5] Vassilakis, S; Petropoulos, G; Sokovic, M; Vaxevanidis, NM; Ntziantzias, I. Macro-geometric and Surface Roughness Quality of TRIP Steel Sheets Processed by Abrasive Waterjet Machining. *Proc. 9th International Conference on Management of Innovative Technologies-MIT 2007*, 8th - 10th October, 2007, Fiesa, Slovenia, pp. 196-203.

[6] Benardos, PG; Vosniakos, GC. Predicting surface roughness in machining: a review. *International Journal of Machine Tools and Manufacture*, 2003 43, 833-844.

[7] Petropoulos, G; Mata, F; Davim, JP. Statistical study of surface roughness in turning of peek composites. *Materials and Design*, 2008 29, 218-223.

[8] Jesuthanam, CP; Kumanan, S; Asokan, P. Surface roughness prediction using hybrid neural networks. *Machining Science and Technology*, 2007 11, 271-286.

[9] Markopoulos, A; Vaxevanidis, NM; Petropoulos, G; Manolakos, DE. Artificial neural network modeling of surface finish in Electro-discharge machining of tool steels.

Proceedings of ESDA2006: 8th Biennial ASME Conference on Engineering Systems Design and Analysis, July 4-7, 2006, Torino, Italy. (paper ESDA2006-95609).

[10] Markopoulos, A; Manolakos, DE; Vaxevanidis, NM. Artificial Neural Network Models for the Prediction of Surface Roughness in Electrical Discharge Machining. *Journal of Intelligent Manufacturing*, 2008 19, 283-292.

[11] Snoeys, R; Staelens, F; Dekeyser, W. Current Trends in Non-Conventional Material Removal Processes. *Annals of the CIRP*, 1986 35, 467-480.

[12] Mc Geough JA. *Advanced Methods of Machining*, Chapman and Hall, London, 1988.

[13] Ma, C; Deam, RT. A correlation for predicting the kerf profile from abrasive water jet cutting. *Experimental Thermal and Fluid Science*, 2006 30, 337-343.

[14] Hoogstrate, AM; van Luttervelt, CA. Opportunities in Abrasive Water-Jet Machining. *Annals of the CIRP*, 1997 46, 697-714.

[15] Kovacevic, R., Hashish, M., Mohan, R., Ramulu, M., Kim, T.J., Geskin, E.S. State of the art of research and development in abrasive waterjet machining. *Journal of Manufacturing Science and Engineering (Transactions of the ASME)*, 1997 119, 776-785.

[16] Wang, J; Wong, WCK. A study of abrasive waterjet cutting of metallic coated sheet steels. *International Journal of Machine Tools and Manufacture*, 1999 39, 855–870.

[17] Davalo, E; Naim, P; Rawsthorne, A. *Neural networks*, Macmillan Education, London, 1991.

[18] Fausset, LV. *Fundamentals of neural networks: Architectures, algorithms and applications*, Prentice Hall, New Jersey, 1994.

[19] Haykin, S. *Neural networks, a comprehensive foundation*, Prentice Hall, New Jersey, 1999.

[20] Feng, C-XJ; Yu, Z-G; Kusiak, A. Selection and validation of predictive regression and neural network models based on designed experiments. *IIIE Transactions*, 2006 38, 13-23.

[21] Tsai, K-M; Wang, PJ. Predictions on surface finish in electrical discharge machining based upon neural network models. *International Journal of Machine Tools and Manufacture*, 2001 41, 1385-1403.

[22] Caydas, U; Hascalık, A. A study on surface roughness in abrasive waterjet machining process using artificial neural networks and regression analysis method. *Journal of Materials Processing Technology*, 2008 202, 574–582.

[23] Deb, S; Dixit, US. Intelligent Machining: Computational Methods and Optimization, (ch. 12). In Davim, JP, editor. *Machining - Fundamentals and Recent Advances.* Springer; 2008; 329- 358.

[24] Parikh PJ; Lam, SS. Parameter estimation for abrasive water jet machining process using neural networks. *International Journal of Advanced Manufacturing Technology*, 2009 40, 497–502.

[25] Yang, L; Song, J; Hu, B. Neural network parametric modelling of abrasive waterjet cutting quality. *International Journal of Abrasive Technology*, 2007 1, 198-207.

[26] Petropoulos, G; Tsolakis, N; Vaxevanidis, N; Antoniadis, A. Topographic description of abrasive waterjet machined surfaces. *Proc. 2nd European Conference on Tribology - ECOTRIB 2009*, 7-10 June 2009, Pisa, Italy, pp. 309-314.

[27] Demuth, H; Beale, M. *Neural networks toolbox for use with Matlab, User's guide.* 2001.

[28] Karbadakis, NA; Mekras, ND; Kontovazenitis, PV; Petropoulos, GP; Vaxevanidis, NM. Implementation of a knowledge based system for modeling non-conventional machining processes. *Nonconventional Technologies Review*, 2007 1, 43-48.

In: Artificial Intelligence in Manufacturing Research
Editor: J. Paulo Davim
ISBN 978-1-62618-557-9

Chapter 6

MULTI-OBJECTIVE OPTIMISATION OF CUTTING PARAMETERS FOR DRILLING ALUMINIUM AA1050

***Ramón Quiza*[a] *and J. Paulo Davim*[*,b]**

[a]Department of Mechanical Engineering, University of Matanzas, Matanzas, Cuba
[b]Department of Mechanical Engineering, University of Aveiro Campus Santiago, Aveiro, Portugal

ABSTRACT

This paper presents a multi-objective approach to the optimisation of drilling process of aluminium AA1050, considering two important, conflicting objectives: material removal rate and obtained surface roughness. From experimental data were developed statistical models for the significant variables involved in the cutting process. To carry out the optimisation process, two methods were used: the numerical method GRG2 and a multi-objective genetic algorithm. Both methods were compared taking into account their respective advantages and drawbacks. The paper remarks the conveniences of the a posteriori approach in the optimisation and decision-making process.

NOMENCLATURE

d	Hole diameter [mm];
F_F	Feed force [N];
f	Feed [mm/rev];
M	Material removal rate [mm^3/s];
M_T	Cutting torque [N·m];
n	Spindle speed [rpm];
P	Cutting power [kW];

[*] Corresponding Author, E-mail: pdavim@ua.pt

P_{MOT}	Motor power [kW];
Ra	Surface roughness [μm];
V	Cutting speed [m/min];
η	Transmission efficiency;

1. Introduction

In general, machining aluminium alloys required cutting fluids because the tendency of the chip to stick to the rake face of the cutting tool. However, the appropriate selection of the cutting fluid and lubrication technique is a hard task because the great number of parameters involved. Recently Mendes et al. [1] investigated the performance of cutting fluids when machining aluminium alloys. Subsequently, Davim et al [2] presented an experimental study on drill of aluminium (AA1050) under dry, minimum quantity of lubricant and flood-lubricated conditions. The results of the tests show it is possible to obtain similar performances to flood-lubricated conditions by using minimum quantity of lubricant.

Optimal selection of cutting parameters is a key issue in any machining process. However, it remains been a difficult work [3] in spite of much work has been done in this field and several optimisation approach have been proposed [4].

Although the common practice in cutting parameters optimisation is considering the whole problem as single-objective, the complex nature of the cutting processes actually makes the problem multi-objective [5]. Some multi-objective approaches have been reported in cutting parameters optimisation [6-8] but mainly they use a priori techniques, where the decision maker combines the different objectives into a scalar cost function. This actually simplifies the multi-objective problem, turning it into a single-objective one [9].

On the other hand, in the a posteriori techniques, the decision-maker is presented with a set of non-dominated optimal candidate solutions and chooses from that set. These solutions are optimal in the wide sense that no other solution in the search space is superior to them when all optimisation objectives are simultaneously considered [10]. They are also known as Pareto-optimal solutions. The main advantage of this approach is to allow selecting the most proper solution under different conditions, without running again the optimisation process.

In obtaining the Pareto-optimal solutions, classical optimisation methods (weighted sum methods, goal programming, min–max methods, etc.) are not efficient, because they cannot find multiple solutions in a single run, thereby requiring them to be applied as many times as the number of desired Pareto-optimal solutions [11].

On the contrary, studies on evolutionary algorithms have shown that these methods can be efficiently used to eliminate most of the above-mentioned difficulties of classical methods [12]. Nevertheless, there is not any certainty that evolutionary methods can obtain actual optimal values, for each specific problem [13].

Quiza and co-workers [14] carried out an a posteriori multi-objective optimisation for drilling laminate composite materials, by using a genetic algorithm, considering productivity and surface quality as optimisation targets. Outputs show the advantages of the a posteriori approach, but they are not compared with classical (numerical) methods.

In this paper an aluminium AA1050 drilling process is modelled and optimised, considering two objectives: productivity (material removal rate) and surface quality (roughness). The a posteriori approach is used and two methods, numeric and evolutive, are employed to find the Pareto-optimal solutions. Outcomes form both methods are compared.

2. Experimental Study

In order to achieve the objective of this experimental work, Aluminium AA1050 (aluminium with max. 0.4%Fe, 0.25%Si, 0.05% Cu, 0.05%Mg, 0.05%Mn, 0.05%V, 0.05% Zn and 0.03%Ti) was tested. A machining center "VCE500 MIKRON" with 11kW spindle power and maximum spindle speed of 7500 rpm were used to perform the cut holes (diameter 5 mm) in 15 mm thick aluminium discs under flood-lubricated conditions.. A helical K10 drill (R415.5-0500-30) was manufactured according DIN6537 by Sandvik®. The drill was a point angle 140°, 10% cobalt grade and 28 mm of flute length. A Kistler® piezoelectric dynamometer 9272 with a load amplifier was used to acquire the cutting torque, MT; and the feed force, FF. Data acquisitions were made through piezoelectric dynamometer by interface RS-232 to load amplifier and PC using the appropriate software Dynoware Kistler®.

The surface roughness was evaluated (Ra according to ISO 4287/1) with a Hommeltester T1000 profilometer. A full factorial design was carried out for the two independent variables: cutting speed, V, and feed, f. Experimental data is shown in Table 1.

Table 1. Experimental data

No.	V [m/min]	f [rev/min]	n [rpm]	M_T [N·m]	F_F [N]	Ra [μm]
1	60	0.15	3820	1.02	833	1.80
2	60	0.20	3820	1.27	1113	2.09
3	60	0.25	3820	1.54	1373	2.13
4	75	0.15	4775	1.00	897	1.78
5	75	0.20	4775	1.22	1154	1.93
6	75	0.25	4775	1.45	1367	2.00
7	90	0.15	5730	0.98	826	1.64
8	90	0.20	5730	1.18	1110	1.87
9	90	0.25	5730	1.32	1206	1.75

3. Modelling

Models were obtained from the experimental data by using multiple regression analyses. For every studied variable several models were tried and analysed. They were compared by considering their correlation coefficients and the variance of the model. Furthermore, the influence of each independent variable was analysed by means of the corresponding t-Student tests, and the lack of autocorrelation was indicated by the outcomes of the Durbin-Watson (DW) tests. Finally, the most convenient model was selected for each dependent variable.

For the surface roughness, the most suitable model is the quadratic form:

$$Ra = -0.29 - 45.33 f^2 - 0.04347\, fV + 23.59 f \tag{1}$$

The R-squared statistic for the above-mentioned model is 95.6 % and the mean absolute error is 0.0276. Since the p-value in the ANOVA table (see Table 2) is less than 0.01, there is a statistically significant relationship between the variables at the 99% confidence level. The DW statistic has a value of 2.504 and an associated p-value of 0.2209, therefore there is no there is no indication of serial autocorrelation in the residuals. In Figure 1 is shown the graphical representation of the adjusted model for the surface roughness.

The most convenient model for the cutting torque is:

$$M_T = \frac{9.758 f^{0.7053}}{V^{0.2186}} \tag{2}$$

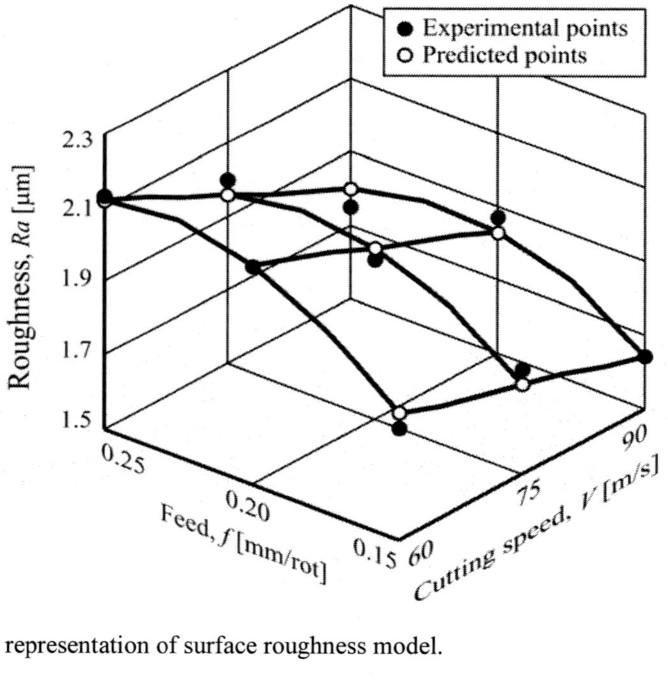

Figure 1. Graphical representation of surface roughness model.

Table 2. ANOVA for the surface roughness model

Source	Sum of Squares	D.F.	Mean Square	F-Ratio	P-Value
Model	0.2046	3	0.0682	36.31	0.0008
Residual	0.0094	5	0.0019	---	---
Total (Corr.)	0.2140	8	---	---	---

Table 3. ANOVA for the cutting torque model

Source	Sum of Squares	D.F.	Mean Square	F-Ratio	P-Value
Model	0.2076	2	0.1038	165.3	0.0000
Residual	0.0038	6	0.0006	---	---
Total (Corr.)	0.2114	8	---	---	---

This model has an R-squared statistic of 98.2 % and a mean absolute error of 0.0165. As can be seen in Table 3, the P-value of the ANOVA is less than 0.01, so there is a statistically significant relationship between the variables at the 99% confidence level. The DW test showed a value of 0.0868, with an associated p-value of 0.060. Since this value is greater than 0.05 there is no indication of serial autocorrelation in the residuals. A graphical representation of the cutting torque model is shown in Figure 2.

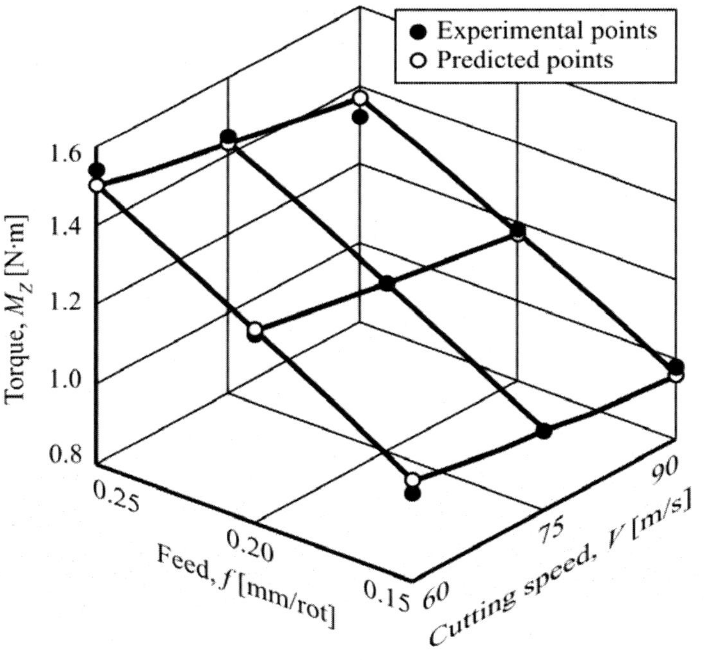

Figure 2. Graphical representation of cutting torque model.

For the feed force, was adjusted the following model:

$$F_F = 171.0 + 5501f - 11.57fV \tag{3}$$

For this model, the R-square statistic is 94.7 % and the mean absolute error 38.29. In Table 4 is shown the ANOVA. Since the p-value is less than 0.01 there is a statistically significant relationship between the variables at the 99% confidence level. Because the value of the DW statistic is 1.450, with an associated p-value of 0.114, there is no indication of serial autocorrelation in the residuals. In Figure 3 is shown the graphical representation of the feed force model.

Table 4. ANOVA for the feed force model

Source	Sum of Squares	D.F.	Mean Square	F-Ratio	P-Value
Model	329551	2	164775	54.28	0.0001
Residual	18213	6	3035	---	---
Total (Corr.)	347764	8	---	---	---

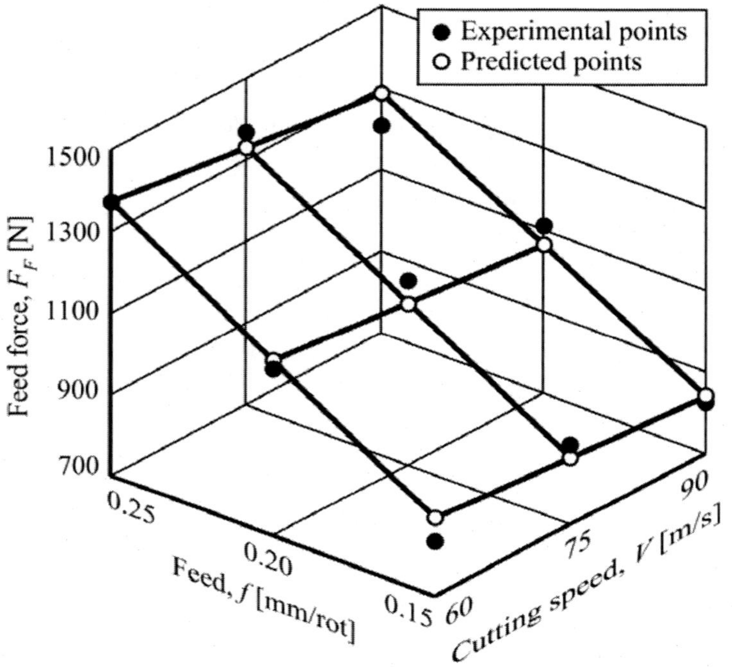

Figure 3. Graphical representation of feed force model.

4. Optimisation

4.1. Formulation of the Optimisation Problem

In the analysed optimisation problem, as decision variables were considered the cutting parameters: feed and speed. Two conflicting objectives were simultaneously optimised. The first is the material removal rate, M, which can be computed as:

$$M = 250dfV \tag{4}$$

This parameter allows evaluating the productivity of the drilling process and is inversely proportional to the machining time. The second objective is the surface roughness, which can be computed by the adjusted model (Eq. 1). It characterises the surface quality of the elaborate hole. As can be noted, the first objective must be maximised since the second one must minimised. Some constraints must be considering in the optimisation process. In the first place, the machining parameters must be into the experimental ranges, in order to guarantee the validity of the previously obtained models.

$$f_{MIN} \leq f \leq f_{MAX} \tag{5a}$$

$$V_{MIN} \leq V \leq V_{MAX} \tag{5b}$$

Moreover, the cutting torque and the feed force must be less or equal to the maximum values allowed by the machine tool.

$$M_T \leq M_{T-MAX} \tag{6a}$$

$$F_F \leq F_{F-MAX} \tag{6b}$$

Furthermore, the cutting power:

$$P = \frac{2\pi M_T n}{6 \times 10^4} \leq \eta P_{MOT} \tag{7}$$

must be satisfied by the machine tool motor, considering the losses in the transmission.

In this problem the following values was considered (see Table 5):

Table 5. Optimisation parameters for the considered problem

Parameter	Value
Hole diameter	5 mm
Maximum allowed cutting torque	1.25 N·m
Maximum allowed feed force	1200 N
Motor power	5.6 kW
Transmission efficiency	75 %

4.2. Optimisation by Numerical Method

In order to obtain a Pareto's front by numerical method, one objective is transformed into constraint. Several values are successively assigned to this objective and the other one is optimised once for every value established for the first one. In this problem, surface roughness was transformed into constraint and material removal rate was kept as objective. To solve the optimisation problem the Generalized Reduced Gradient (GRG2) algorithm was used. In Table 6, there are shown the outcomes of the optimisation process.

Table 6. Outcomes of the optimisation by using the GRG2 method

No.	V [m/min]	f [rev/min]	M_T [N·m]	F_F [N]	n [rpm]	P [kW]	Ra [µm]	M [mm^3/s]
1A	90.0	0.219	1.25	844	5730	0.750	1.85	49264
2A	90.0	0.185	1.11	885	5730	0.666	1.80	41706
3A	90.0	0.171	1.05	934	5730	0.630	1.75	38508
4A	90.0	0.160	1.00	996	5730	0.601	1.70	36088
5A	90.0	0.151	0.96	1148	5730	0.577	1.65	34059

4.3. Optimisation by Using Genetic Algorithm

The other considered approach carries out the optimisation process by using a micro genetic algorithm. This genetic algorithm was originally proposed by Coello and Toscano [15] and it was used by Quiza and co-workers to solve optimisation problems in machining processes [14, 16].

Table 7. Outcomes of the optimisation by using the micro-GA

No.	V [m/min]	f [rev/min]	M_T [N·m]	F_F [N]	n [rpm]	P [kW]	Ra [µm]	M [mm^3/s]
1B	89.0	0.217	1.25	1141	5665	0.739	1.86	48275
2B	89.2	0.213	1.23	1123	5676	0.730	1.85	47480
3B	89.3	0.204	1.19	1082	5685	0.709	1.84	45546
4B	89.8	0.198	1.16	1055	5716	0.697	1.83	44446
5B	89.3	0.191	1.14	1024	5682	0.677	1.82	42618
6B	89.2	0.184	1.11	993	5677	0.659	1.80	41017
7B	89.8	0.180	1.09	974	5715	0.652	1.79	40398
8B	90.0	0.176	1.07	956	5729	0.643	1.77	39599
9B	89.7	0.174	1.06	948	5709	0.636	1.76	39009
10B	89.7	0.168	1.04	921	5713	0.621	1.74	37690
11B	89.2	0.164	1.02	904	5679	0.607	1.72	36574
12B	88.9	0.160	1.00	887	5657	0.595	1.71	35542
13B	89.2	0.158	0.99	877	5679	0.592	1.69	35236
14B	89.2	0.155	0.98	864	5676	0.584	1.68	34549
15B	89.1	0.153	0.97	855	5675	0.578	1.67	34094
16B	88.7	0.150	0.96	842	5649	0.568	1.65	33275

For the considered application, the algorithm's parameters were selected as follow: static population size: 500, dynamic population size: 20; chromosome length: 64 (32 for each decision variable), maximum epochs count: 100; maximum evolutionary periods count: 50; mutation likelihood: 0.01 and maximum Paretian size: 25. After carrying out the optimisation process, there were obtained 10 near-optimal solutions, which are shown in Table 7.

5. DISCUSSION

To compare outcomes from both approaches, a Pareto's fronts graph was constructed (see Figure 4). In this graph, non-dominated solutions are arranged in a coordinated plane whose axes are the optimisation objectives.

As can be seen, the solutions contained in the front obtained by the numerical method are better than contained in the front obtained by micro genetic algorithm. The term "better" means that for every solution belong to the numerical obtained front there is a point in the genetic algorithm obtained front that dominates it. Moreover, the genetic front has a higher density of points, allowing selecting the most proper solution from a larger set. Naturally, executing new runs of the optimisation process must increase the amount of points in the numerical front. Nevertheless, this must be done by try-and-error and can take a great amount of time and personal effort.

The main drawback of the genetic algorithm approach is its elevated computational cost, determined by the high amount of evaluations of the objective functions that must be executed. However, this fact is widely paid back by a practically total elimination of human interaction in the optimisation process. It should be noted that the try-and-error process, used in the numerical approach, is not only highly time consuming but also very tedious and ineffective from the human user point of view.

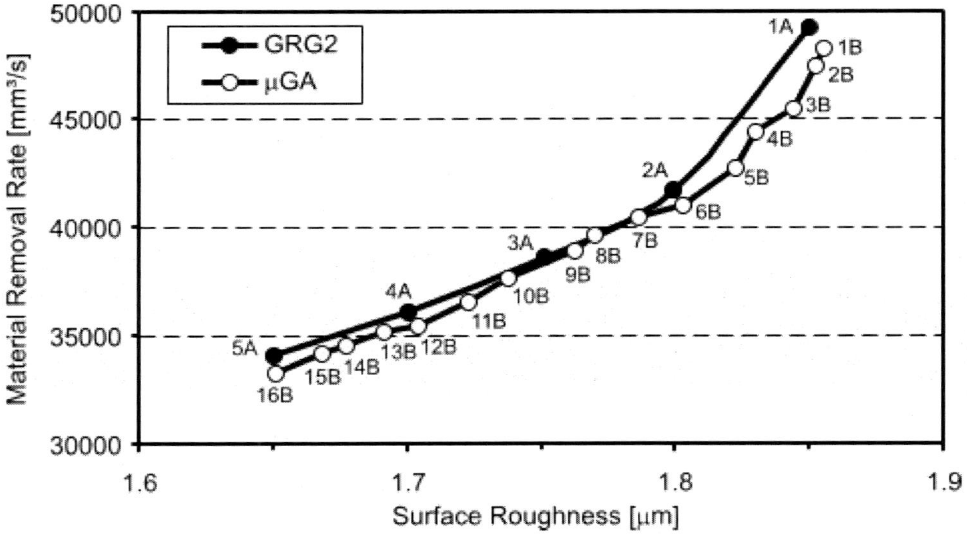

Figure 4. Paretos's fronts.

CONCLUSION

As conclusions of the work can be remarked the convenience of the use of evolutionary algorithms in multi-objective optimisation of cutting processes, instead its high computational cost. Finally, the advantage of the *a posteriori* approach in the cutting parameters optimisation must be remarked. From the Pareto's front graph a flexible decision-making can be done, considering different technological requirement in the elaborated holes. By using a pre-elaborated Pareto's front, the selection of the most convenient cutting parameters, for an established surface roughness can be done without carry out the optimisation process once again. For example, if the only important criterion is productivity, the extreme point 1B must be chosen. This point has a poor surface quality but corresponds to the highest material removal rate. On the contrary, if the surface quality must be keep as low as possible, the other extreme point, 16B, is the most appropriated instead its low productivity. In an intermediate case, where a hole must be drilled with a surface roughness of Ra 1,75 μm, point 10B must be selected.

ACKNOWLEDGMENTS

The authors acknowledge to the MSc A. Festas and P. Reis as well as the Mechanical Engineers C. Peixoto and R. Gomes for their participation in the experimental work carried in University of Aveiro.

REFERENCES

[1] Mendes, O. C.; Ávila, R. F.; Abrão, A. M.; Reis, P.; Davim, J. P. *Ind. Lubric. Trib.* 2006, 58 (5), 260-268.
[2] Davim, J. P.; Sreejith, P. S.; Gomes, R.; Peixoto, C. *Proc. IMechE Part B: J. Eng. Manuf.* 2006, 220, 1605-1611.
[3] Jain, R. K.; Jain, V. K. *J. Mater. Process. Technol.* 2000, 108 (1), 62-67.
[4] Mukherjee, I.; Ray, P. K. *Comput. Ind. Eng.* 2006, 50 (1), 15-34.
[5] Yih-Fong, T.; Fu-Chen, C. *Int. J. Machin. Machinab. Mat.* 2006, 1 (1), 76-93.
[6] Lee, B. Y.; Tarng, Y. S.; Lii, H. R. *J. Mater. Process. Technol.* 2000, 105 (1), 1-6.
[7] Zuperl, U.; Cus, F. *Robot. Comput. Int. Manuf.* 2003, 19 (1-2), 189-199.
[8] Cus, F.; Balic, J. *Robot. Comput. Int. Manuf.* 2003, 19 (1-2), 113-121.
[9] Van Veldhuizen, D. A.; Lamont, G. B. *Evol. Comput.*, 2000, 8 (2), 125-147.
[10] Abbass, H. A.; Sarker, R. A. In: *Congress on Evolutionary Computation*, Piscataway, NJ (U.S.A.), 2001; pp 971-978.
[11] Coello, C. A. *IEEE Comput. Intell. Mag.* 2006, 1 (1), 28-36.
[12] Soodamani, R.; Liu, Z. Q. *Int. J. Approx. Reason.* 2000, 23(2), 85-109.
[13] Zitzler, E.; Laumanns, M.; Bleuler, S. In: *Metaheuristics for Multiobjective Optimisation*; Gandibleux, X.; Sevaux, M.; Sörensen, K.; T'kindt, V. Eds.; Springer-Verlag: Berlín (Alemania), 2004; pp 3-37.
[14] Quiza, R.; Reis, P.; Davim, J. P. *Compos. Sci. Technol.* 2006, 66 (15), 3083-3088.
[15] Coello, C. A.; Toscano, G. In: *First International Conference on Evolutionary Multi-Criterion Optimization*. Zitzler, E. Ed.; Springer-Verlag: New York (U.S.A.), 2001; pp 126-140.
[16] Quiza, R.; Rivas, M.; Alfonso, E. *Eng. Appl. Artif. Intell.* 2006, 19 (2), 127-133.

In: Artificial Intelligence in Manufacturing Research
Editor: J. Paulo Davim
ISBN 978-1-62618-557-9

Chapter 7

APPLICATION OF FUZZY LOGIC IN MANUFACTURING: A STUDY ON MODELING OF CUTTING FORCE IN TURNING GFRP COMPOSITES

K. Palanikumar* and J. Paulo Davim
Sri Sairam Institute of Technology,
Sai Leo Nagar, Chennai, India
Department of Mechanical Engineering, University of Aveiro
Campus Santiago, Aveiro, Portugal

ABSTRACT

Artificial intelligence (AI) tools are playing an important role in manufacturing engineering. They are used for modeling, control and optimization of parameters in manufacturing. Fuzzy logic is one of the important tools in artificial intelligence and is used in varieties of manufacturing applications. This chapter discusses the application of fuzzy logic for modeling process parameters for cutting force in turning of GFRP composites. Taguchi method in design of experiments principle is used for conducting the experiments. An L_{27} (3^{13}) orthogonal array is used for the investigation. The cutting parameters selected were: cutting speed, feed and depth of cut. Fuzzy rules were developed for correlating the cutting parameters with cutting force in turning of GFRP composites. The model predicted values and measured values are fairly close to each other. The confirmation test results proved the fact that the developed models are effectively representing the cutting force in turning GFRP composites.

1. INTRODUCTION

Artificial intelligence (AI) is the study of ideas that enables computers to be intelligent. AI systems have a significant impact on the automation of engineering manufacture. This will provide intelligent interfaces to sophisticated engineering analysis automate the use of

* Corresponding author, E-mail: palanikumar_k@yahoo.com

manufacturing data base and provide engineering assistance to handle varieties of engineering tasks. AI is used for modeling, control and optimization of parameters in manufacturing. Fuzzy logic is one of the important tool in artificial intelligence, which can be used for modeling, analysis and optimization of manufacturing processes. The fuzzy set theory initiated by Zadeh [1] is a mathematical theory of inexact reasoning that allows modeling of the reasoning process in human linguistic terms. This theory proved to be an effective means for dealing with objectives that are linguistically specified. Linguistic term such as low, medium and high may be defined by fuzzy sets [2]. Fuzzy logic has been applied successfully for many manufacturing processes especially in machining processes. AI techniques are insensitive to the noise produced in the machining data due to variations in parameters such as material properties, temperature, and cutting geometry, the same are being used in modeling of machining processes in the present day research. Suleyman Yaldiz et al. [3] have evaluated the cutting force in turning. They have compared the experimental results obtained by designed dynamometer to fuzzy model for predicting cutting force. Surface roughness prediction for turning was carried out by Yue Jiao et al. [4]. They have used Fuzzy adaptive network for modeling the machining process Fang [5] has used fuzzy logic diagnosis for machining condition monitoring of finish turning process. Palanikumar et al. [6] have used fuzzy logic for optimization of machining parameters in machining GFRP composites. This method is highly useful when the process is complex and uncertain in nature. The works carried out by the researchers indicated that fuzzy logic can be applied for manufacturing processes. In the present chapter, application of fuzzy logic for the modeling of cutting force in turning GFRP composites is introduced. Fuzzy rule based modeling technique is used for the present investigation.

Glass fiber reinforced composite finds many applications including construction industries, marine, electrical, chemical, pharmaceutical, transportation, sport goods manufacturing industries etc. Composites materials exhibit good resistance to corrosion and wear. The tailorability of composites for specific applications has been one of its greater advantages and also one of the more perplexing challenges to adopting are alternative to metallic materials in many situations. Even though GFRP parts may be produced by molding processes, they require further machining to facilitate dimensional control for easy assembly and control of surface quality for functional aspects. Machining of FRP parts includes turning, drilling, milling, grinding and the like. The machining of fiber reinforced composite materials are different from the traditional engineering materials. Fibre delamination, fuzziness, debonding of fiber and matrix materials, excessive wear on the cutting tool are some of the problems which encountered during the machining of FRPs [7]. Over the last several years, the phenomenon of machining of FRP composites has been analyzed by researchers. Everstine and Rogers [8] proposed an analytical theory of machining FRPs. In their classical study, they prepared a theory related to plane deformation of incompressible composites reinforced by strong parallel fibers.

In machining process, cutting force plays an important role in deciding the surface and power requirement for cutting. Generally the machining force increases with the feed rate and decreases with the cutting velocity [9]. The machining force evaluation in cutting GFRP composites has been carried out by Paulo Davim et al. [10,11]. They have analyzed the cutting force in drilling and milling operation using cemented carbide drill bits and end mill cutter. The influence of cutting force in turning GFRP composites is carried out by Sang-Ook et al. [12]. It was found that the single crystal diamond tool was performed better than both poly crystal diamond and cubic boron nitride tools in turning of GFRP composites.

Eventhough many attempts have been made to analyze the cutting force in turning of GFRP composites, there is no systematic study has been reported. The present chapter investigates the effect of different machining parameters on cutting force in machining of GFRP composites. The experiments are conducted on a lathe using Poly Crystalline Diamond (PCD) cutting tools. The experiments are conducted based on L_{27} orthogonal array with prefixing cutting parameters. The machining parameters considered for the experiments are cutting speed, feed, and depth of cut. Fuzzy rule based modeling technique is used for modeling the machining parameters in machining of GFRP composites. The results indicate that fuzzy logic modeling technique can be successfully applied for the prediction of cutting force in turning of GFRP composites.

2. EXPERIMENTAL DETAILS

The work piece material used for the present investigation is glass fiber reinforced polymer composite pipes consisting of E-glass fibers and polyester resin. The pipes are manufactured through filament winding process. The fiber volume fraction for the composites is 55%. The machining operation considered is turning and the machine used for the machining is Bharat make all geared lathe of model Nagmati 175 with a spindle speed range of 54 - 1200 rpm and power of 2.25 kW. The cutting tool used for the machining is PCD tool which is shown in Figure 1. The experiments for the investigation are conducted by using Taguchi's orthogonal array [13], the suitable array selected for the investigation was L_{27}, which needs 27 runs and has 26 DOFs. It can conduct three levels of parameters. To check the degrees of freedom (DOFs) in the experimental design, for the three-level test, the three main factors take 6 (3*(3-1)) DOFs. The DOF for three second-order interactions is 12 (3*((3-1)*(3-1))) and the total DOFs required is 18. As per Taguchi's experimental design method, the total DOFs of selected orthogonal array must be greater than or equal to the total DOFs required for the experiment and hence L_{27} orthogonal array has been selected [14].

Figure 1. Tool holder and cutting tool used for the investigation.

The cutting parameters selected for the present investigation are cutting speed (V), depth of cut (d) and feed (f). The cutting parameters used and their levels considered are presented in Table 1. The three level L_{27} orthogonal array is shown in Table 2, where the numbers 1, 2, and 3 stand for the levels of the factors. This array specifies 27 experimental run and have 13 columns. The assignment of cutting parameters and interactions to columns is presented in Table 2.

Table 1. Experimental factors and levels

Factor	Notation	Unit	Factor levels		
			1	2	3
Cutting speed	V	m/min	82	126	170
Depth of cut	d	mm	0.5	1.0	1.5
Feed	f	mm/rev	0.096	0.143	0.191

Table 2. Experimental Layout Using L_{27} Orthogonal Array

Column numbers													
	1	2	3	4	5	6	7	8	9	10	11	12	13
Trial No.	V	d	$V*d$	$V*d$	f	$V*f$	$V*f$	$d*f$	--	--	$d*f$	--	--
1	1	1	1	1	1	1	1	1	1	1	1	1	1
2	1	1	1	1	2	2	2	2	2	2	2	2	2
3	1	1	1	1	3	3	3	3	3	3	3	3	3
4	1	2	2	2	1	1	1	2	2	2	3	3	3
5	1	2	2	2	2	2	2	3	3	3	1	1	1
6	1	2	2	2	3	3	3	1	1	1	2	2	2
7	1	3	3	3	1	1	1	3	3	3	2	2	2
8	1	3	3	3	2	2	2	1	1	1	3	3	3
9	1	3	3	3	3	3	3	2	2	2	1	1	1
10	2	1	2	3	1	2	3	1	2	3	1	2	3
11	2	1	2	3	2	3	1	2	3	1	2	3	1
12	2	1	2	3	3	1	2	3	1	2	3	1	2
13	2	2	3	1	1	2	3	2	3	1	3	1	2
14	2	2	3	1	2	3	1	3	1	2	1	2	3
15	2	2	3	1	3	1	2	1	2	3	2	3	1
16	2	3	1	2	1	2	3	3	1	2	2	3	1
17	2	3	1	2	2	3	1	1	2	3	3	1	2
18	2	3	1	2	3	1	2	2	3	1	1	2	3
19	3	1	3	2	1	3	2	1	3	2	1	3	2
20	3	1	3	2	2	1	3	2	1	3	2	1	3
21	3	1	3	2	3	2	1	3	2	1	3	2	1
22	3	2	1	3	1	3	2	2	1	3	3	2	1
23	3	2	1	3	2	1	3	3	2	1	1	3	2
24	3	2	1	3	3	2	1	1	3	2	2	1	3
25	3	3	2	1	1	3	2	3	2	1	2	1	3
26	3	3	2	1	2	1	3	1	3	2	3	2	1
27	3	3	2	1	3	2	1	2	1	3	1	3	2

The cutting force developed during machining is measured by using a Kistler quartz 3-component dynamometer type 9257B. The arrangement used for measuring cutting force is given in Figure 2.

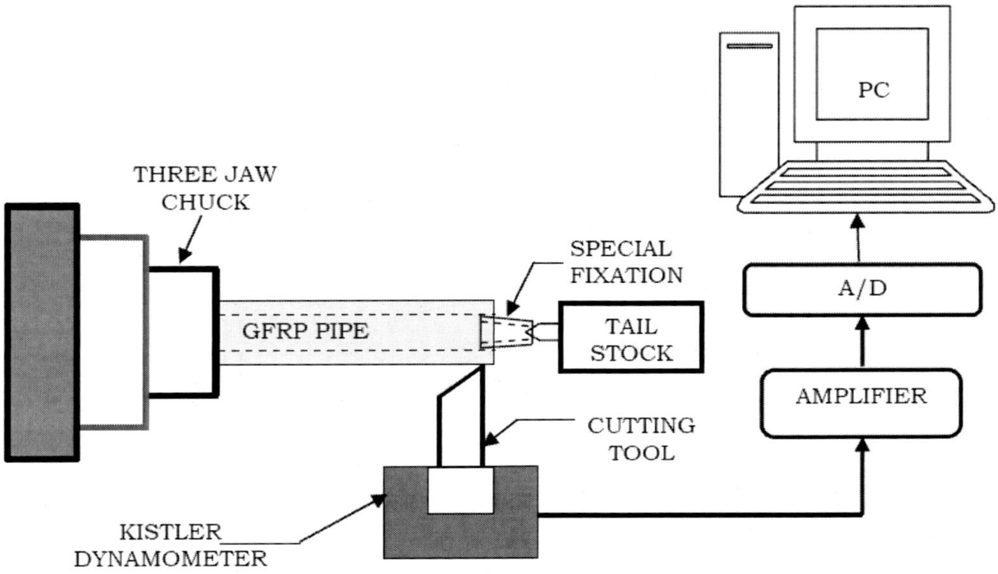

Figure 2. Schematic Diagram of Experimental Setup.

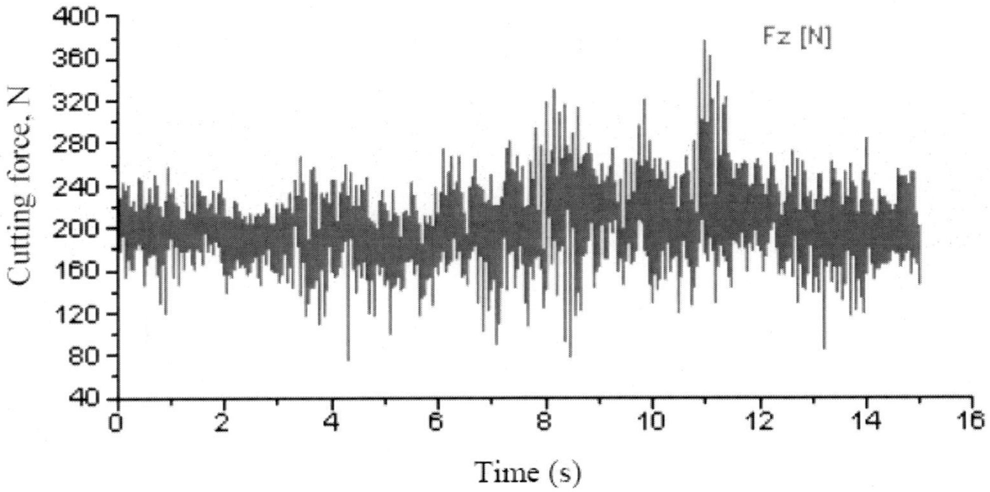

Figure 3. Typical cutting force observed in machining of GFRP composites.

The experiments are repeated for three times and the average values are used for further analysis. The dynamometer measures the active cutting force regardless of its application point. Both the average value of the force and the dynamic force increase may be measured. The usable frequency range depends mainly on the resonance frequency of the entire measuring rig. The force to be measured is introduced via a top plate and distributed between four 3-component force sensors arranged between the base and top plates. Each of the sensors has three pairs of quartz plates, one sensitive to pressure in the z-direction and the other two to shear in the x and y directions respectively. The measurement is virtually without displacement. The dynamometer is connected to a 3-channel charge amplifier type 5807A through a connecting cable type 1687B5 which in turn is connected to the PC by a 37-pin cable from the A/D board. The dynamometer is calibrated for the cutting force in the range from 0 to 1000 N. The typical cutting force observed from the Kistler dynamometer is presented in Figure 3.

The experimental condition used and the experimental results are presented in Table 3.

Table 3. Experimental results

S.No	Cutting speed (*V*), m/min	Depth of cut (*d*), mm	Feed rate (f), mm/rev.	Cutting force(F_c), N
1	82	0.5	0.096	178.5
2	82	0.5	0.143	201.2
3	82	0.5	0.191	231.5
4	82	1	0.096	192
5	82	1	0.143	207.8
6	82	1	0.191	242.4
7	82	1.5	0.096	212.5
8	82	1.5	0.143	213
9	82	1.5	0.191	251.4
10	126	0.5	0.096	171.2
11	126	0.5	0.143	190.8
12	126	0.5	0.191	225
13	126	1	0.096	187.9
14	126	1	0.143	201.7
15	126	1	0.191	236.6
16	126	1.5	0.096	172.3
17	126	1.5	0.143	212.3
18	126	1.5	0.191	241.8
19	170	0.5	0.096	173.5
20	170	0.5	0.143	192.6
21	170	0.5	0.191	221.6
22	170	1	0.096	179.7
23	170	1	0.143	203.6
24	170	1	0.191	234.6
25	170	1.5	0.096	191.6
26	170	1.5	0.143	208.4
27	170	1.5	0.191	241.5

3. Fuzzy Rule Based Model

Fuzzy logic is based on imprecision, specifically, the way people make decisions on imprecise and non-numerical information. Fuzzy logic mimics the way humans make decisions using linguistic reasoning. Fuzzy logic modeling is based on mathematical theory combining multi-valued logic, probability theory and Artificial Intelligence methods and can be used to tackle complex problems. Fuzzy modeling is based on fuzzy set theory and the linguistic statements expressed mathematically which corresponds to the analysis of a human expert. Fuzzy systems base their decisions on inputs and outputs in the form of linguistic variables. The variables are tested with IF-THEN rules, which produce one or more responses depending on which rules are asserted. The response of each rule is weighed according to the degree of membership of its inputs and the centroid of the responses is calculated to generate the appropriate output [15 -17].

3.1. Fuzzy Set Theory

The fuzzy set theory, introduced by Zadeh in 1965 in his seminar paper entitled 'Fuzzy Sets' to deal with domains of pattern recognition, communication of information and abstraction [15]. A fuzzy set as the name implies, is a set without a crisp boundary, that is, the transition from "belong to a set" to "not belong to a set" is gradual and this smooth transition is characterized by membership functions that give fuzzy sets flexibility in modeling commonly used linguistic expressions such as " the water is hot " or the " the temperature is high " [18].

If X is a collection of objects denoted generally by x, then a fuzzy set A in X is defined as a set of ordered pairs:

$$A = \{(x, \mu_A(x)) \mid x \in X\}, \tag{1}$$

where $\mu_A(x)$ is called the membership function for the fuzzy set A. This function maps each element of X to a membership grade or value between 0 and 1 and $(x, \mu_A(x))$ is a singleton. Usually X is referred to as the universe of discourse or simply the universe and it may consist of discrete objects or continuous space [18]. Another common way of representing a fuzzy set is

$$A = \bigcup_{x_i \in X} \mu_A(x_i)/x_i \tag{2}$$

Here the fuzzy set A is the collection or union of all $\mu_A(x_i)/x_i$.

If set A is a crisp set, it is determined by the membership function that precisely identifies the boundaries of the set (a,b) on the universe of discourse. If it is a fuzzy set, it is determined by the membership function that shows the distribution of degrees of relevance across the universe of discourse [19].

Let X denote a universal set. Then the membership function μ_A by which a fuzzy set A is usually defined has the form:

$$\mu_A : \mathrm{X} \rightarrow [0,1] \tag{3}$$

For example, we can define a possible membership function for the fuzzy set of real numbers close to 0 as follows.

$$\mu_A(x) = \frac{1}{1+10x^2} \tag{4}$$

Given a crisp universal set X, let $\overline{P}(X)$ denote the set of all fuzzy subsets of X and let $\overline{P}^k(X)$ be defined recursively by the equation

$$\overline{P}^k(X) = \overline{P}(\overline{P}^{k-1}(X)) \tag{5}$$

for all integers $k \geq 2$. Then fuzzy sets of level k are formally defined by membership functions of the form

$$\mu_A : \overline{P}^{k-1}(X) \rightarrow [0,1] \tag{6}$$

A fuzzy set A is defined on universal set that is finite and countable is written as

$$A = \sum_{i=1}^{n} \mu_i / x_i \tag{7}$$

$$A = \int_X \mu_A(x) / x \tag{8}$$

3.2. The Fuzzy Inference System

The fuzzy inference system is a popular computing framework based on the concepts of fuzzy set theory, fuzzy IF – THEN rules, and fuzzy reasoning. It has been used in a wide variety of fields, such as automatic control, data classification, decision analysis, expert system, and time series prediction, robotics and pattern recognition. It is a knowledge based system that converts linguistic variables into its knowledgebase using fuzzy rules and membership functions. Because of its multidisciplinary nature, the fuzzy inference system is known by numerous other names, such as fuzzy – rule based system, fuzzy expert system, fuzzy model, fuzzy associative memory, fuzzy logic controller and simply fuzzy system. The basic structure of a fuzzy inference system consists of three conceptual components: a rule base, which contains a selection of fuzzy rules; a data base (or dictionary), which defines the membership functions used in the fuzzy rules; and a reasoning mechanism, which performs the inference procedure upon rules and given facts to derive a reasonable output or conclusion [14-20]. The concept of fuzzy reasoning for three-input-one-output fuzzy logic unit is described as follows. The fuzzy rule base consists of a group of IF- THEN statements with three inputs, x_1, x_2 and x_3 and one output y, i.e.

Rule 1: if x_1 is A_1 and x_2 is B_1 and x_3 is C_1 then y is D_1 else
Rule 2: if x_1 is A_2 and x_2 is B_2 and x_3 is C_2 then y is D_2 else
....
....
Rule n: if x_1 is A_n and x_2 is B_n and x_3 is C_n then y is D_n. (9)

A_i, B_i, C_i and D_i are fuzzy subsets defined by the corresponding membership functions, i.e., μ_{A_i}, μ_{B_i} μ_{C_i} and μ_{D_i}. 27 fuzzy rules were developed. By taking the max-min compositional operation, the fuzzy reasoning of these rules yields a fuzzy output. Suppose that x_1, x_2 and x_3 are the three input values of the fuzzy logic unit, the membership function of the output of fuzzy reasoning can be expressed as [20]:

$$\mu_{D_o}(y)=[\mu_{A_1}(x_1)\wedge\mu_{B_1}(x_2)\wedge\mu_{C_1}(x_3)\wedge\mu_{D_1}(y)\vee..\mu_{A_n}(x_1)\wedge\mu_{B_n}(x_2)\wedge\mu_{C_n}(x_3)\wedge\mu_{D_n}(y)] \tag{10}$$

where $\wedge$ is the minimum operation and $\vee$ is the maximum operation. Finally, a defuzzification method is used to transform the fuzzy output into a non-fuzzy value y_0. The membership functions can be of different forms like triangular, trapezoidal, Gaussian, sigmoid, etc. The triangular shaped membership function for input is specified by three parameters {a, b, c} as follows:

$$\text{triangle}(x; a, b, c) = \begin{cases} 0, & x \le a. \\ \dfrac{x-a}{b-a}, & a \le x \le b. \\ \dfrac{c-x}{c-b}, & b \le x \le c. \\ 0, & c \le x. \end{cases} \tag{11}$$

By using min and max, an alternate expression for the proceeding equation is

$$trimf(x;a,b,c) = \max\left(\min\left(\frac{x-a}{b-a},\frac{c-x}{c-b}\right),0\right) \tag{12}$$

where a, b, c stands for the triangular fuzzy triplet and it determines the x coordinates of the three corners of the underlying triangular membership function. For example, the functions considered for cutting speed is LOW, MEDIUM and HIGH as shown in Figure 4. Similarly for other input parameters, the membership functions have been fixed and are shown in Figures 5 and 6. The output responses of the fuzzy process can be viewed only in fuzzy values and they have to be defuzzified. The membership function used for the output response cutting force is presented in Figure 7.

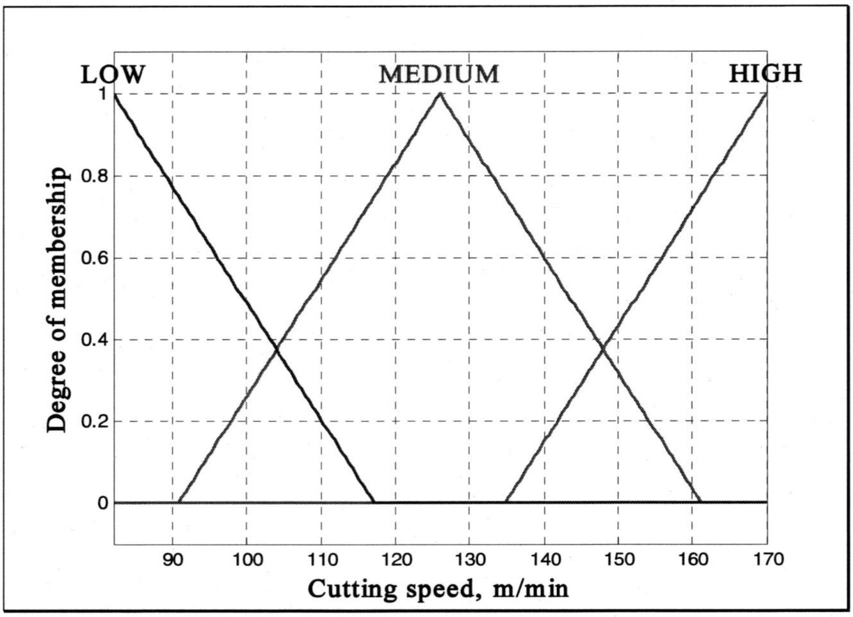

Figure 4. Membership function for cutting speed.

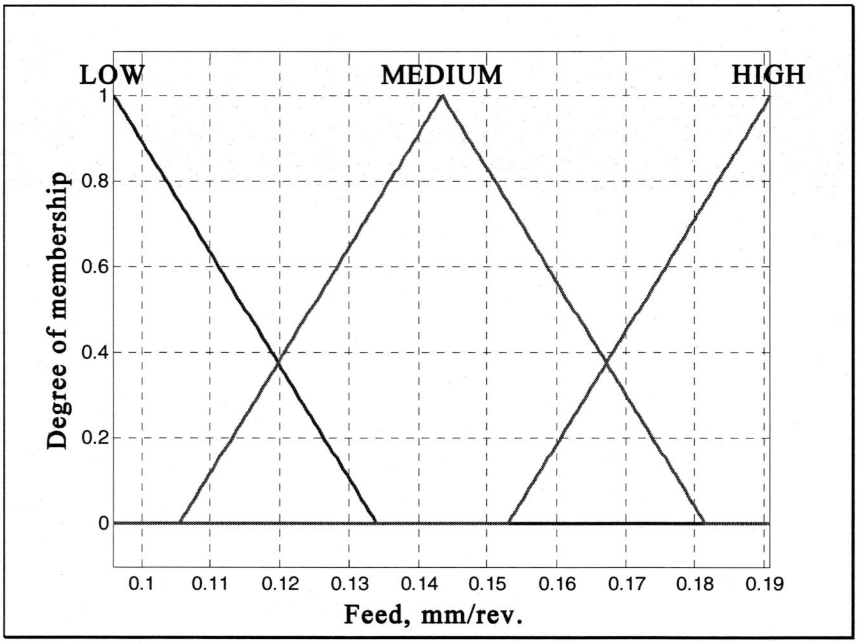

Figure 5. Membership function for feed.

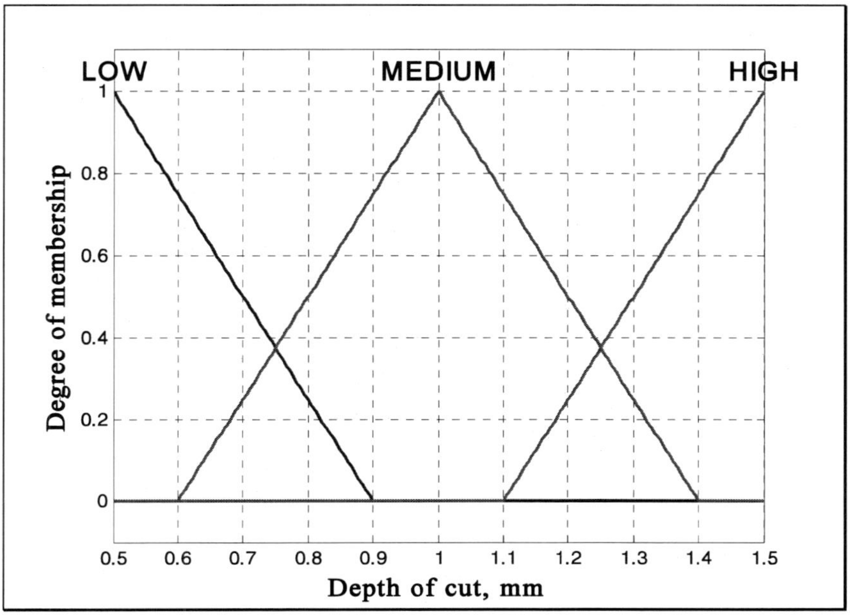

Figure 6. Membership function for depth of cut.

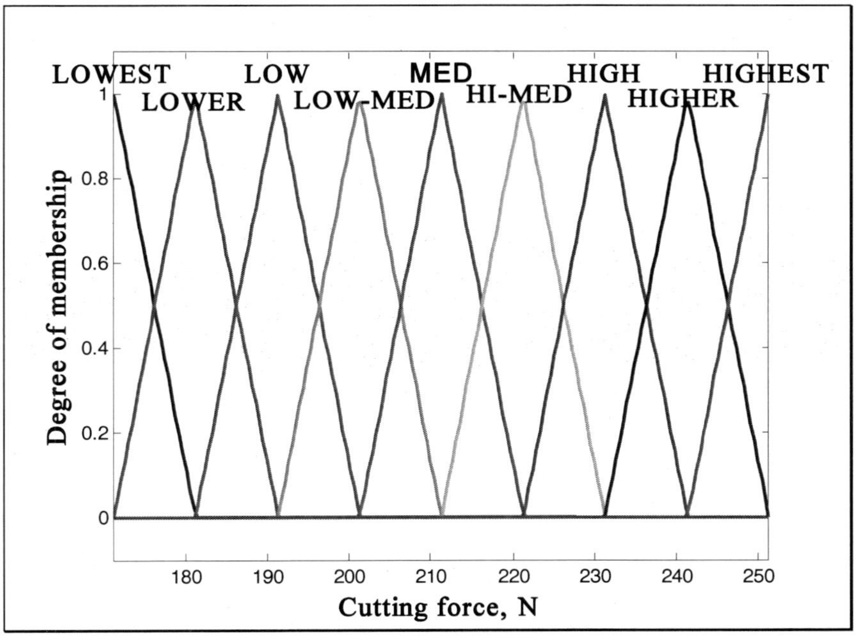

Figure 7. Membership function for output response cutting force in machining GFRP composites.

The linguistic forms used for the input parameters and output response cutting force in machining of GFRP composites with respect to experiment number is presented in Table 4.

Table 4. Linguistic form used for cutting parameters and output response cutting force

S.No	Cutting speed (*V*), m/min	Depth of cut (*d*), mm	Feed rate (f), mm/rev.	Cutting force (F_c), N
1	LOW	LOW	LOW	LOWEST
2	LOW	LOW	MEDIUM	LOW-MED
3	LOW	LOW	HIGH	HIGH
4	LOW	MEDIUM	LOW	LOW
5	LOW	MEDIUM	MEDIUM	MEDIUM
6	LOW	MEDIUM	HIGH	HIGHER
7	LOW	HIGH	LOW	MEDIUM
8	LOW	HIGH	MEDIUM	MEDIUM
9	LOW	HIGH	HIGH	HIGHEST
10	MEDIUM	LOW	LOW	LOWEST
11	MEDIUM	LOW	MEDIUM	LOW
12	MEDIUM	LOW	HIGH	HIGH
13	MEDIUM	MEDIUM	LOW	LOWER
14	MEDIUM	MEDIUM	MEDIUM	LOW-MED
15	MEDIUM	MEDIUM	HIGH	HIGHER
16	MEDIUM	HIGH	LOW	LOWEST
17	MEDIUM	HIGH	MEDIUM	MEDIUM
18	MEDIUM	HIGH	HIGH	HIGHER
19	HIGH	LOW	LOW	LOWEST
20	HIGH	LOW	MEDIUM	LOW
21	HIGH	LOW	HIGH	HIGH-MED
22	HIGH	MEDIUM	LOW	LOWEST
23	HIGH	MEDIUM	MEDIUM	LOW-MED
24	HIGH	MEDIUM	HIGH	HIGHER
25	HIGH	HIGH	LOW	LOW
26	HIGH	HIGH	MEDIUM	MEDIUM
27	HIGH	HIGH	HIGH	HIGHER

In the present study, the centroid defuzzification method has been selected, as it produces the centre of area of the possibility distribution of the inferenced output and is a more frequently used defuzzification method calculating the centroid of the area under the membership function

$$y_0 = \frac{\sum y\mu_{D_0}(y)}{\sum \mu_{D_0}(y)} \tag{13}$$

The non-fuzzy value y_0 gives the output value in numerical form. For example the value of cutting force at a cutting condition of cutting speed of 126 m/min, depth of cut of 1mm and feed of 0.143 mm/rev., is obtained as 201 N. Matlab® fuzzy logic tool box is used for this calculation and calculated result is shown in Figure 8. In the above example single rule is applied. Also for different input values multiple rules can be applied. In the example Mamdani max-min approach is used as an inference engine.

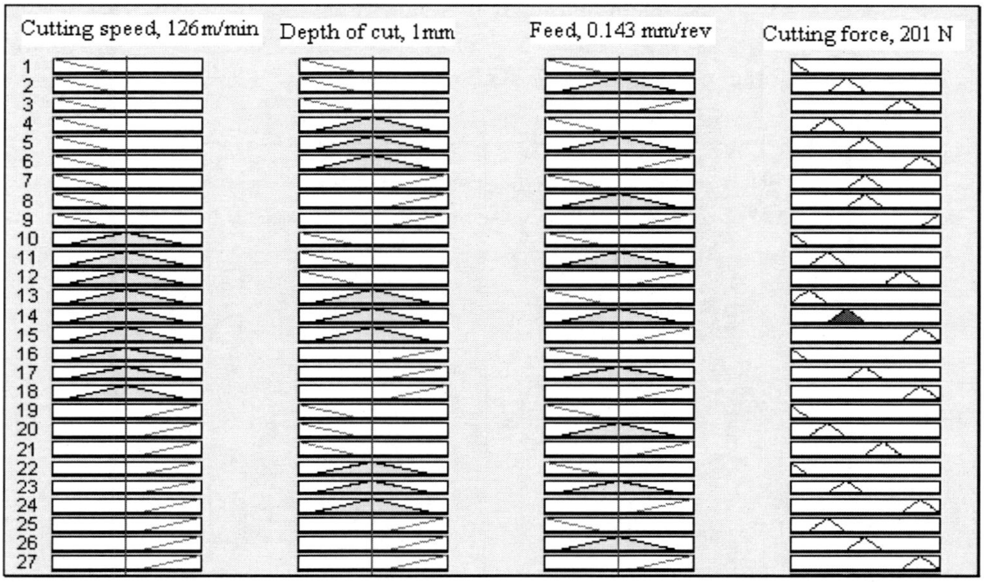

Figure 8. Cutting force observed using fuzzy logic.

4. Results and Discussions

The study on cutting force in turning of GFRP composites is one of the important areas of research. To maintain the proper surface and reducing the power requirement for cutting minimization of cutting force is required. The minimization of cutting force is the serious task because fibre reinforced composite materials is made of two different materials like the hard fibers and soft matrix materials. The property of the materials may not be uniform throughout. Machining of glass fiber reinforced composite materials poses big problems for the manufacturing engineers. Fibre pull-out, delamination, matrix crazing, high tool wear are some of the problems encountered while machining GFRPs. The cutting parameters such as cutting speed, depth of cut and feed are playing a vital role in deciding the cutting force in turning of GFRP composites. Also the cutting force on the machining of composites depends on material properties, orientation of fibers in the matrix, type of weave, bond strength, etc.

For analyzing the cutting force in turning of GFRP composites, mathematical relations are required. In this investigation, Fuzzy rule based modeling technique is used for the modeling of machining parameters in machining of GFRP composites with respect to the cutting force. Fuzzy logic models are one of the important tools in the artificial intelligence and can be effectively used for the modeling of manufacturing processes such as turning.

Figure 9 shows the correlation between the cutting force observed during machining and fuzzy logic model. The adequacy of the developed model is verified through R-Sq (R^2) value, which is presented in the figure. The quantity R-Sq is called as coefficient of determination and is used to judge the adequacy of the models developed. In the present case the coefficient of determination is 0.9848 which shows high correlation that exists between the fuzzy logic model developed and experimental results and hence fuzzy logic can be effectively used for the prediction of cutting force in turning of GFRP composites. Further the comparison of results obtained from the experiment and fuzzy model is plotted in Figure 10, in which all the points are coincide with each other and hence fuzzy logic can be effectively utilized for the prediction of cutting force in machining of GFRP composites.

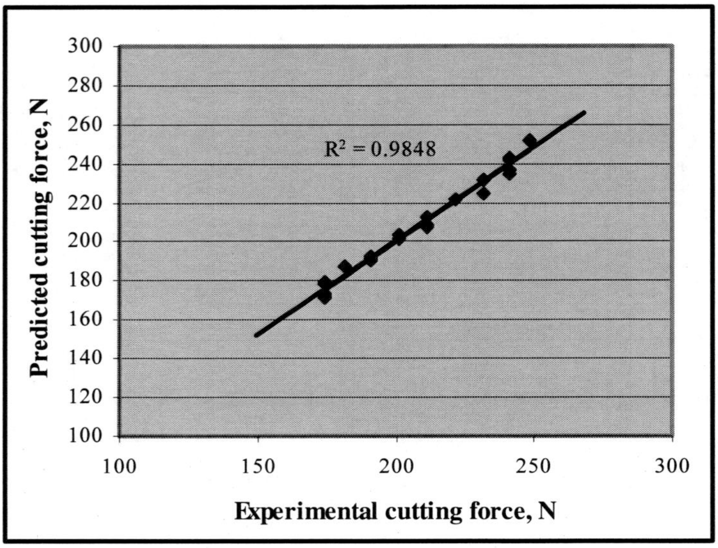

Figure 9. Correlation graph for cutting force.

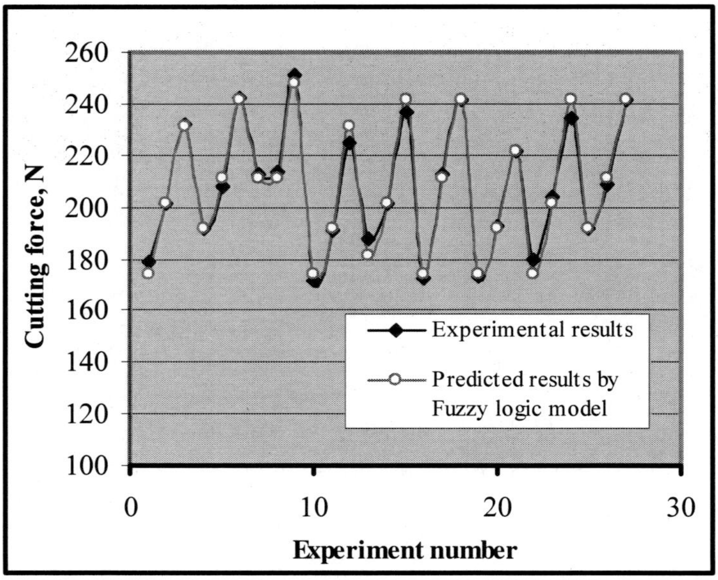

Figure 10. Comparison of experimental result and fuzzy output for cutting force.

The effect of cutting parameters such as cutting speed, depth of cut and feed can be analyzed through three-dimensional surface plots. Surface plots are helping to visualize the response surface. This plot represents the functional relationship between the response and the experimental factors. The response surface plots also help to visualize how the response reacts to changes in the experimental factors. The surface plot shows only two factors at a time, and the extra factor should be kept at a constant level.

Figure 11 shows the surface plot of cutting force with variables cutting speed and depth of cut while keeping the feed at constant middle level. The figure indicates that the increase of cutting speed reduce the cutting force in turning of GFRP composites. The reason being, at high cutting speed, the removal of glass fibers in the GFRP composites is easy and it produces low cutting force. The increase in depth of cut increases the cutting force in turning of GFRP composites. The increase in cutting force increases the load on the cutting tool and it leads to the increased cutting force.

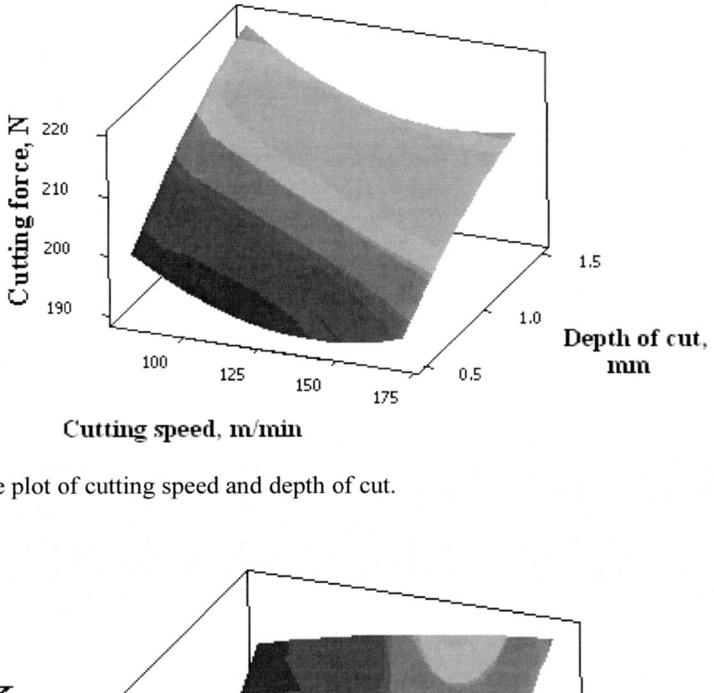

Figure 11. Surface plot of cutting speed and depth of cut.

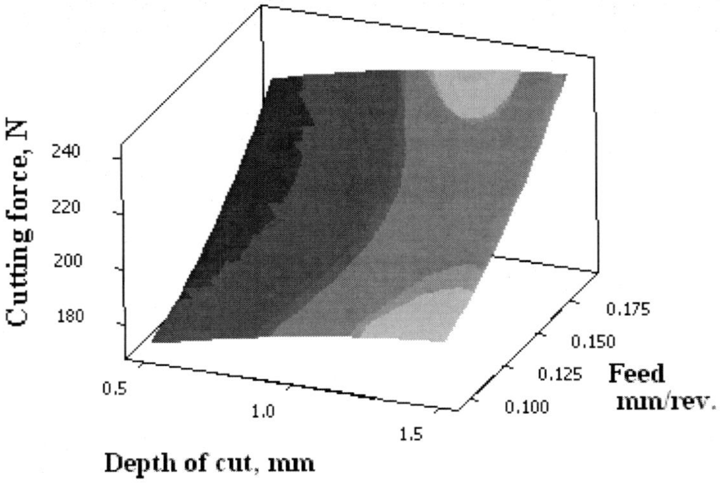

Figure 12. Surface plot of depth of cut and feed.

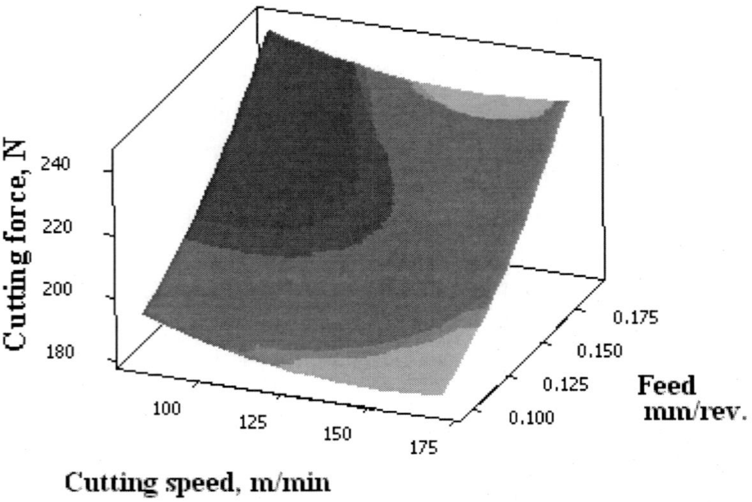

Figure 13. Surface plot of cutting speed and feed.

The surface plot of varying the variables depth of cut and feed is depicted in Figure 12. The figure indicates that the increase of feed increases the cutting force abruptly. The reason being the increase of feed rate increases the tooth load on the cutting tool and also produces chatter or vibration in cutting of GFRP composites which in turn increases the cutting force. The effect of cutting speed and feed on cutting force is presented in Figure 13. The figures indicated that the increase of feed and depth of cut increases the cutting force in turning of GFRP composites. The magnitude of cutting force variation is high while increasing the feed compared to the depth of cut. From the graphs, one can infer that high cutting speed, low depth of cut and low feed are preferred for cutting the GFRP composites.

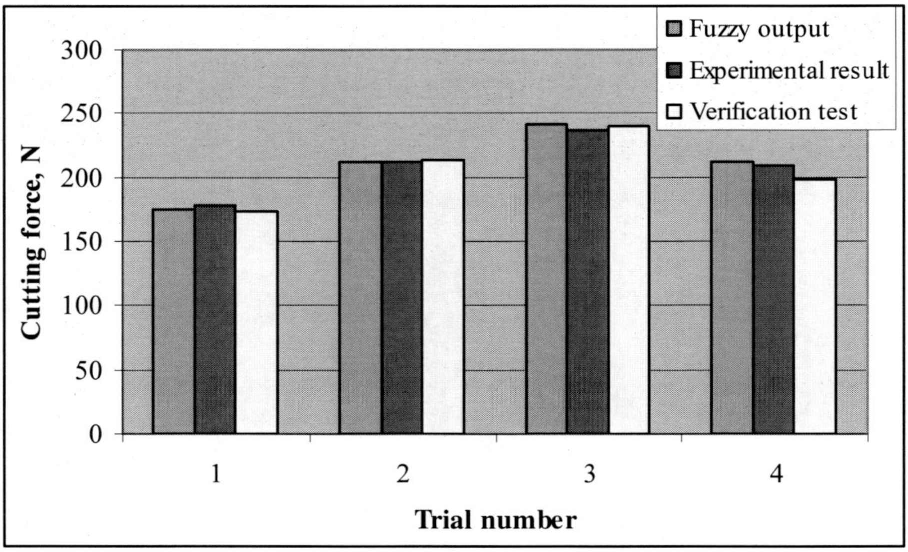

Figure 14. Verification test result for cutting force.

For verification of fuzzy logic model and experimental results, verification tests were carried out. The verification tests were carried out as per the corresponding experimental numbers 1, 7, 15 and 26. The verification test results for cutting force in turning are presented in Figure 14. From the figures it can be asserted that the variation between experimental results and the model are within the limit and are very close to each other and hence fuzzy logic model can be effectively used for the prediction of cutting force parameters in turning of GFRP composites.

CONCLUSIONS

Experiments are conducted for modeling the parameters which affect the cutting force in turning of GFRP composites. Taguchi's experimental design has been used for experimentation. Fuzzy rule based modeling technique is employed for modeling the machining parameters. Based on the experimental results and data analysis, the following conclusions can be drawn.

1. Fuzzy rule based modeling technique is used for predicting the cutting force in turning of GFRP composites.
2. The model for cutting force is well correlated with experimental results. The results indicated that artificial intelligence assisted tool can be well suited for the prediction of cutting force in turning GFRP composites.
3. In machining of GFRP composites, the increase in cutting speed reduces the cutting force, whereas the increase of feed increases the cutting force.
4. In machining of GFRP composites feed is the main parameter which affects the cutting force in machining of GFRP composites.
5. The confirmation experiment reveals that the developed model can be effectively used for predicting the cutting force in turning of GFRP composites.
6. The artificial intelligence tool used for the manufacturing process is simple and can be used as an on-line monitoring tool, if proper equipments are used.

REFERENCES

[1] Zadeh L.A. (1965), *Fuzzy sets, Information and control*, 8:338-353.

[2] Hasmi K., Graham I.D and Mills B, (2000), Fuzzy logic based data selection for the drilling process, *Journal of Materials Processing Technology* 108: 55-61.

[3] Suleyman Yaldiz, Faruk Unsacar and Haci Saglam (2006), Comparison of experimental results obtained by designed dynamometer to fuzzy model for predicting cutting forces in turning, *Materials and Design*, 27: 1139-1147.

[4] Yue Jiao, Shuting Lei, Pei Z.J, and Lee E.S, (2004), Fuzzy adaptive networks in machining process modeling: surface roughness prediction for turning operations, *International Journal of Machine Tools and Manufacture*, 44:1643-1651.

[5] Fang, X.D (1995). Expert system support fuzzy diagnosis of finishturning process states. *International Journal of Machine Tools and Manufacture 35* (6): 913–924.

[6] Palanikumar, K.; Karunamoorthy, L.; Karthikeyan, R.; Latha, B (2006). Optimization of machining parameters in turning GFRP composites using a carbide (K10) tool based on the Taguchi method with fuzzy logics. *Metals and Materials 12/6*: 483–491.

[7] Santhanakrishnan, G., Krishnamoorthy, R. and Malhotra, S. K. (1989). High Speed Steel Tool Wear Studies in Machining of Glass Fiber-reinforced Plastics, *Wear*, 132: 327–336.

[8] Evestine, G. C. and Rogers, T. G. (1971). A Theory of Machining of Fiber-reinforced Materials, *J. Composite Materials*, 5: 94–106.

[9] Arno kopf, StefenFeistritzer, Klaus Udier (2006). Diamond coated cutting tools for machining of non-ferrous metals and fiber reinforced polymers, *International Journal of Refreactory metals and hard materials*, 24: 354-359.

[10] Paulo Davim J. Pedro Reis. Conceicao A.C (2004/). Experimental study of drilling glass fiber reinforced plastics (GFRP) manufactured by hand lay up, *Computer Science and Technology*, 64(2): 289 – 297.

[11] Paulo Davim J. Pedro Reis. Conceicao Antonio C, (2004). A study on milling of glass fiber reinforced plastics manufactured by hand-lay up using statistical analysis (ANOVA)', *Composite Structures*, 64: 493-500.

[12] Sang-Ook An. Eun-Sang Lee Sang-Lai Noh, (1997), A study on the cutting characteristics of glass fibre reinforced plastics with respect to toll materials and geometries' *Journal of Materials Processing Technology*, 68: 60-67.

[13] Ross, P. J. (1989). *Orthogonal Array Selection and Utilization*, Taguchi Techniques for Quality Engineering, McGraw-Hill, New York.

[14] Palanikumar, K, (2006). Cutting parameters optimization for surface roughness in machining of GFRP composites using Taguchi's method, *Journal of Reinforced plastics and composites,* 25(16): 1739-1751.

[15] Nandi A.K. and Pratihar D.K, (2004). An expert system based on FBFN using a GA to predict surface finish in ultra-precision turning", *Journal of Materials Processing Technology*, 155-157: 1150-1156.

[16] Oguzhan Yilmaz, Omer Eyercioglu, Nabil N.Z. Gindy (2006), A user-friendly fuzzy-based system for the selection of elsctro discharge machining process parameters, *Journal of Materials Processing Technology* , 172: 363-371.

[17] Palanikumar. K (2009), Surface roughness model for machining glass fiber reinforced plastics by PCD tool using fuzzy logics, *Journal of Reinforced Plastics and Composites,* Jun 2009; DOI. 0731684408092009v1.

[18] Jang, J.S.r. Sun C.T, Mizutani E., Neuro – Fuzzy and soft computing – A Computational approach to learning and machine intelligence, *Pearson Education*, 2005.

[19] Riza C. Berkan, Sheldon L. .Trubatch , Fuzzy systems design principles –Building fuzzy IF-THEN Rule bases, *IEEE Press,* Standard Publishers Distributors, 2000.

[20] Lin J.L., Wang K.S., Yan B.H., Tarng Y.S. (2000), Optimization of the electrical discharge machining process based on the Taguchi method with fuzzy logic, *J. Mater Process*. Technol. 102; 48-55.

In: Artificial Intelligence in Manufacturing Research
Editor: J. Paulo Davim
ISBN 978-1-62618-557-9

Chapter 8

FLANK WEAR DETECTION WITH AE SIGNAL AND FNN DURING TURNING OF AL/15 VOL% SIC-MMC

Alakesh Manna*

Department of Mechanical Engineering, Punjab Engineering College, Deemed University, Chandigarh, India

ABSTRACT

This paper describe predictive machining approach with fuzzy neural network (FNN) modeling of the cutting tool flank wear in order to estimate the performance of CCGX 09 T3 04 Al-H 10 insert during turning of Al/15 vol%SiC-MMC. In the present work a new approach for cutting tool wear detection with cutting conditions estimated wear through acoustic emission (AE) signal is presented. The measured tool wear and estimated tool wear by conditions monitoring and detected signals are compared and graphically analyzed. Investigated results prove that the new method of FNN is reliable and appropriate to control and monitor the cutting tool wear.

1. INTRODUCTION

In modern high-tech industries, manufacturing of various complex components machining processes plays an important role which involving metal cutting from a significant part of most manufacturing task. To ensure the require degree of reliability in the final product, there is a need to maintain high degree of quality control. Automation in the metal cutting process is reliable methods of monitoring the cutting tool wear and tool failure. Sensors and automatic control can monitor the variation of cutting parameters during metal cutting. To fulfill the object several sensing technique have been developed in the recent year for estimating and automatic control the variation of cutting parameter during metal cutting.

* Corresponding author, mail: kgpmanna@rediffmail.com

Some of these techniques have provided quiet useful under laboratory conditions; few of them have been discussed in a review article (Tlusty and Andrews, 1983). One of the important techniques is the acoustic emission can be used to measure the cutting tool wear. Very beginning of tool failure is also indicated by some of these parameters as sudden changes in their characteristics the sensor is expected to be highly sensitive to the parameters, it is to monitor and it should be at the same time sensitive to the others parameters of the metal cutting process as well as to any varying conditions not under process control. It is essential that a sensor is used for monitoring the progressive tool wear in machining, it should have minimum sensitivity to process parameter such as cutting speed, feed and depth of cut, On the other hand sensor whose output is to be used for monitoring cutting speed and also an input in automatic control of the machining process should not be affected by the condition of tool wear. Acoustic emission (AE) is an important method can meet the ideal requirement to monitor the cutting tool wear during metal cutting.

The frequency distribution pattern of force signals is influenced by cutting conditions (Rao and Shin, 1999). Not only frequency distribution pattern of force signal but also AErms, the total energy of forces are influenced by the cutting conditions, tool wear, chip fracture, cutting edge deterioration (Chung Choo and Saini 2000; Usui and Hirata, 1978, Usui et. al. 1978). The AE can be used effectively to detect the tool condition as the frequency of the AE signal is much higher than that of machine vibrations and other noises which eliminates the interference in cutting (Iwata and Moriwaki, 1997; Dornfeld, 1990). In turning operation, with in the range of input variables e.g. cutting speed 20-35 m/min, feed rate 0.075 – 0.125 mm/rev and depth of cut 0.4-0.8 mm the neural network comes ahead of the design of experiment in nearness of the predictions of the experimental values of flank wear (Chaudhury and Bartarya, 2003). Elanayar and Shin (1995) proposed a model, which approximates flank and crater wear propagation and their effects on cutting force by using radial basis function neural networks. The generic approximation capabilities of radial basis function neural networks are used to identify a model and a state estimator is designed based on this identified model. A wide range of tool monitoring techniques utilizing neural networks has been reviewed by Dimla et al. (1997). They concluded that neural networks are adequate for tool condition monitoring. They also pointed out the confusion in the interpretation of TCM techniques in literature as on-line or off-line systems. Ghasempoor et. al. (1999) proposed a tool wear classification and continuous monitoring neural network system for turning by employing recurrent neural network design. In the study of Li et al. (1999), neural network models have also been integrated with analytical models such as Oxley's theory to form a hybrid machining model for the prediction of tool wear and work-piece surface roughness. Neural networks are used to predict difficult-to-model machining characteristic factors. Liu and Altintas (1999) derived an expression to calculate flank wear in terms of cutting force ratio and other machining parameters. The calculated flank wear, force ratio, feed rate and cutting speed are used as an input to a neural network to predict the flank wear in the next step. Chaudhury, S.K; et. al. (1999) used optoelectronic sensor and back propagation of the neural network to predict the tool wear. Very recently an experimental investigation was carried out to predict the ceramic tool wear during machining of D2 AISI steel using neural network by Quiza, Ramon; et. al. (2008) and they concluded that the neural network model has shown better capability to make accurate predictions of tool wear under the conditioned studied.

The different wear on a cutting tool may occur during turning and it is really be very difficult to avoid in practice. However, the flank wear on the cutting tool during turning has the greatest influence on the quality of the machined surface and work-piece dimensional accuracy. Keeping in view, the present research strives to monitor the cutting tool flank wear using AE sensor and based on Fuzzy Neural Net work (FNN) classification.

2. Experimental Condition

Work-piece material: Al/15vol%SiC-MMC is selected as work-piece materials for experimental investigation. Table-1 shows the chemical composition of work-piece used for experimentation. Table 2 represents the parameters and their levels considered for experimental investigation.

Table 1. Chemical composition of Al/SiC-MMC used for experimentation

Types of MMC	Density; APS	%SiC	%Si	%Mg	%Fe	%Cu	%Mn	%Zn	%Ti	Al
LM6Mg 15SiC	2.67g-cm^{-3} 23μm	15.0	12.0	0.50	0.12	0.17	0.11	0.10	0.10	Rem.

Table 2. Physical and Mechanical Properties of Al / SiC – MMC

Material	Density (gm/cm3)	Tensile Strength (Mpa)	Yield Strength (Mpa)	% 0f Elongation	Hardness (BHN)	Modulus of Elasticity (Gpa)
LM6 Mg 15 SiCP	2.67	140	120	1 - 1.5	115	78

Machine Tool; Cutting Tools and Measuring Equipment Used

(i) Machine Tool: Combination Turret Lathe(motor h.p. 7.5, Word Co. England)

Table 3. Details of cutting tool used for experimentation

Cutting tool	Cutting tool specification	Tool material and grade	Rake angle	Clearance angle	Nose radius	Cutting edge angle	Cutting fluid used
T-Max-U Positive rhombic insert	CCGX 09 T3 04 Al-H 10	Uncoated tungsten carbide (WC) (HW-K10)	5^0	7^0	0.4 mm	80^0	Not used

Table 4. Cutting parameters and their levels

Sl. No.	Machining parameters	Level		
		1		3
1	A: Cutting speed, m/min.	40	100	160
2	B: Feed, mm/rev.	0.16	0.32	0.48
3	C: Depth of cut, mm.	0.50	0.75	1.00

(ii) Tool holders Specification: SCLCR 20 20 K 09

(iii) Mitutoyo Shop Microscope with 30X magnification and resolution 1 μm

2.1. Experimental Setup

AE sensor is mounted on the tool shank and the signal output (I_{rms} and $V_{rms,}$ Sensor output) is amplified and converted into the digital form by means of an A/D converter connected in between the PC and Low pass filter. Figure 1 shows the scheme of the experimental setup utilized for experimentation.

As the measure of the AE signal the RMS value of the AE is directly related to the average power dissipation involved in the process, the AE count clearly relates to the occurrence of the discreet events like microscopic slips and fractures in the metals during metal cutting. The AE sensing clearly reveals the information about the hardness and the dislocation densities in the material and therefore has potential to be useful in the identification of individual types of tool wear.

The signal from AE the sensor is first filtered and amplified in a sufficiently wide band to encompass most of the AE spectrum and remove any unwanted noise, especially at low frequency. Assume that AE signal arises mainly from primary contact region between the tool and the chip formed as well as due to the flank wear of the cutting tool and analysis the estimated data.

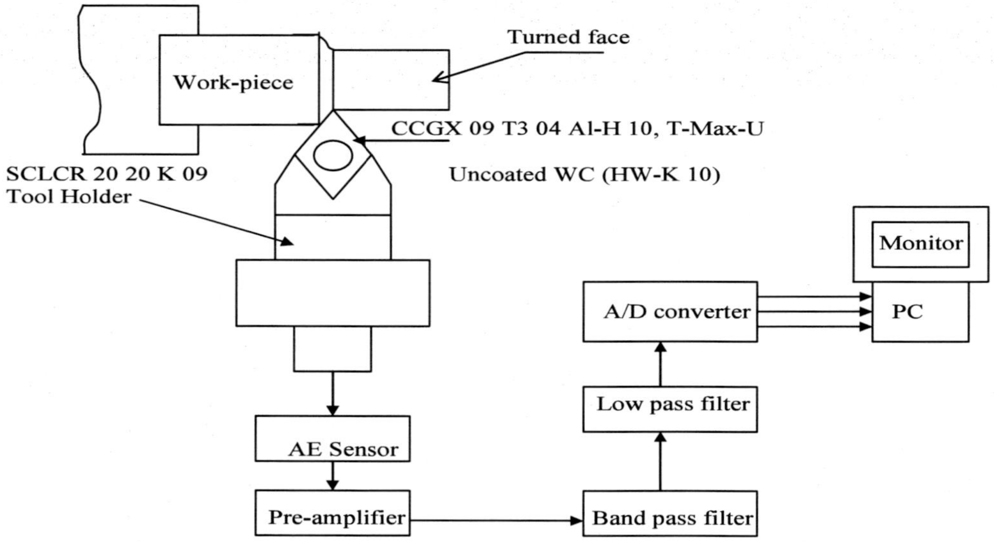

Figure 1. Scheme of the experimental setup.

3. TOOL WEAR DETECTION WITH FNN MODEL

The flow diagram of the user interface is shown in figure 2. The basic function of the developed user interface are promoting the user to enter the necessary information such as turning parameters setting conditions and tool wear size and as well as display the estimated wear to the user. Considering 0.30 mm flank wear is tool life (Bhattacharya, 1984) of the cutting tool insert, flank wear from 0.00 to 0.35 mm was categorized in to seven categories with 0.05 mm increment.

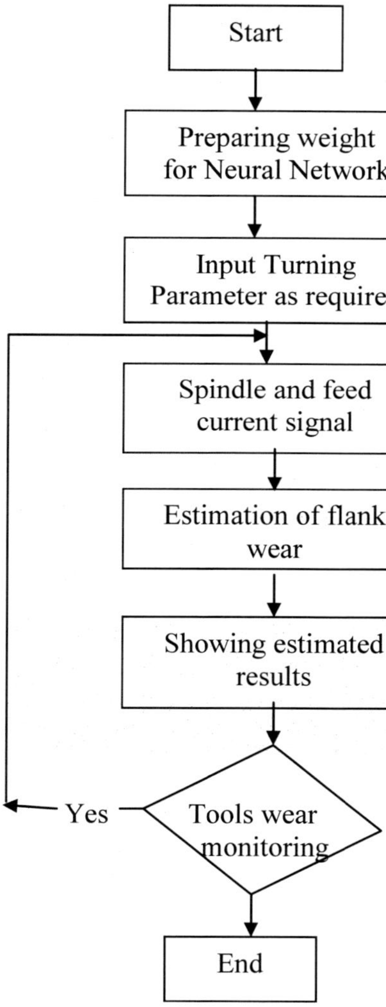

Figure 2. Flow chart for on-line FNN model.

In this study, the flank wear width was categories into Category-A, Category-B, Category-C, Category-D, Category-E, Category-F and Category-G according to the width of the cutting tool flank wear during turning of Al/SiC-MMC using CCGX 09 T3 04 Al-H 10 insert and by utilizing SCLCR 20 20 K 09 cutting tool holder. Table-5 shows the tool wear and their category.

Table 5. Cutting tool flank wear categorization

Category of Flank wear width	A	B	C	D	E	F	G
Flank wear width(mm)	00-0.05	0.05-0.10	0.10-0.15	0.15-0.20	0.20-0.25	0.25-0.30	0.30-0.35

Considering 0.30 mm flank wear is tool life an attempt has been made to develop the model for the current signal. The effect of the cutting parameters e.g. cutting speed (V, m/min), feed rate (f, mm/rev) and depth of cut (d, mm) on the spindle and feed current signal can be expressed in the form of the exponents of the equations as follows

$$I_{spindle} = C_1.V^{n1}.f^{n2}.d^{n3} \quad \text{Eqn.1}$$

$$I_{feed} = C_2.V^{m1}.f^{m2}.d^{m3} \quad \text{Eqn.2}$$

where, $I_{spindle}$ = spindle current (I_S, A), I_{feed} = feed current (I_F, A), C_1 and C_2 are the constant depends on the properties of the work materials and cutting tool and as well as insert geometry; n_1,n_2,n_3 and m_1,m_2,m_3 are the exponential for spindle peak current and feed current respectively. From the above predicted exponents of the equations, it is observed that a unit increase in either the cutting speed, feed rate or depth of cut would have exponential effects of n_1, n_2, n_3 and m_1, m_2, m_3 on the average spindle and feed current signal respectively.

The logarithmic values of the spindle current (I_S, A) can be expressed as follows

$$\begin{pmatrix} I_{S1} \\ I_{S2} \\ I_{S3} \\ I_{S4} \\ I_{S5} \\ I_{S6} \\ I_{S7} \end{pmatrix} = \begin{pmatrix} n_{10}\ n_{11}\ n_{12}\ n_{13} \\ n_{20}\ n_{21}\ n_{22}\ n_{23} \\ n_{30}\ n_{31}\ n_{32}\ n_{33} \\ n_{40}\ n_{41}\ n_{42}\ n_{43} \\ n_{50}\ n_{51}\ n_{52}\ n_{53} \\ n_{60}\ n_{61}\ n_{62}\ n_{63} \\ n_{70}\ n_{71}\ n_{72}\ n_{73} \end{pmatrix} \text{x} \begin{pmatrix} 1 \\ \log V \\ \log f \\ \log d \end{pmatrix} \quad \text{Eqn. 3}$$

The logarithmic values of the feed current (I_F , A) can be expressed as follows

$$\begin{pmatrix} I_{F1} \\ I_{F2} \\ I_{F3} \\ I_{F4} \\ I_{F5} \\ I_{F6} \\ I_{F7} \end{pmatrix} = \begin{pmatrix} m_{10}\ m_{11}\ m_{12}\ m_{13} \\ m_{20}\ m_{21}\ m_{22}\ m_{23} \\ m_{30}\ m_{31}\ m_{32}\ m_{33} \\ m_{40}\ m_{41}\ m_{42}\ m_{43} \\ m_{50}\ m_{51}\ m_{52}\ m_{53} \\ m_{60}\ m_{61}\ m_{62}\ m_{63} \\ m_{70}\ m_{71}\ m_{72}\ m_{73} \end{pmatrix} \text{x} \begin{pmatrix} 1 \\ \log V \\ \log f \\ \log d \end{pmatrix} \quad \text{Eqn. 4}$$

where, I_{S1}, I_{S2} ..... . I_{S7} and I_{F1}, I_{F2}. . . . . .I_{F7} are the algorithmic values of the spindle peak current and feed current respectively. The input variables for this analysis are 1, log V, log f and log d for turning parameters cutting speed (V, m/min), feed rate (f, mm/rev) and depth of cut (d, mm) respectively. Output variables are I_{S1}, I_{S2} ..... . I_{S7} and I_{F1}, I_{F2}. . . . . .I_{F7} for spindle peak current and feed current respectively. Utilizing the above input-output relational matrix the weights of the neural network can be calculated by regression analysis. The matrix equation 3 between the spindle current (I_S, A) and turning parameters, and matrix equation 4 between feed current (I_F, A) and turning parameters can be expressed by the neural network topology is shown in figure 3.

The FNN model can be expressed in four separate heads such as fuzzy logic based tool wear classification, input normalization, neural network base tool wear estimation and fuzzy logic based adjustment of tool wear. As earlier mentioned that the spindle current signal and feed current signal models equation 3 and 4 at the various cutting tool wear can be explained with the help of neural network structure is shown in figure 3. The measure spindle current and feed current are considered as real value and estimated spindle currents for I_{S1}, I_{S2}....I_{S7} and feed current for I_{F1}, I_{F2}....I_{F7} considered as mid value of a bunch of various tool wear classifications. The measure values (i.e. real values) are compared with the estimated values by utilizing fuzzy logic based classification, and the membership degree of the tool wear can be calculated by utilizing the fuzzy curve of membership degree is shown in figure 4. The same method is utilized for both I_S and I_F. Here, three input variables each of which is classified into seven fuzzy sets and an output variables is also classified into seven fuzzy sets and based on the experimental work, forty-nine rules are set. These rules are classified into seven groups corresponding to seven categories of flank wear.

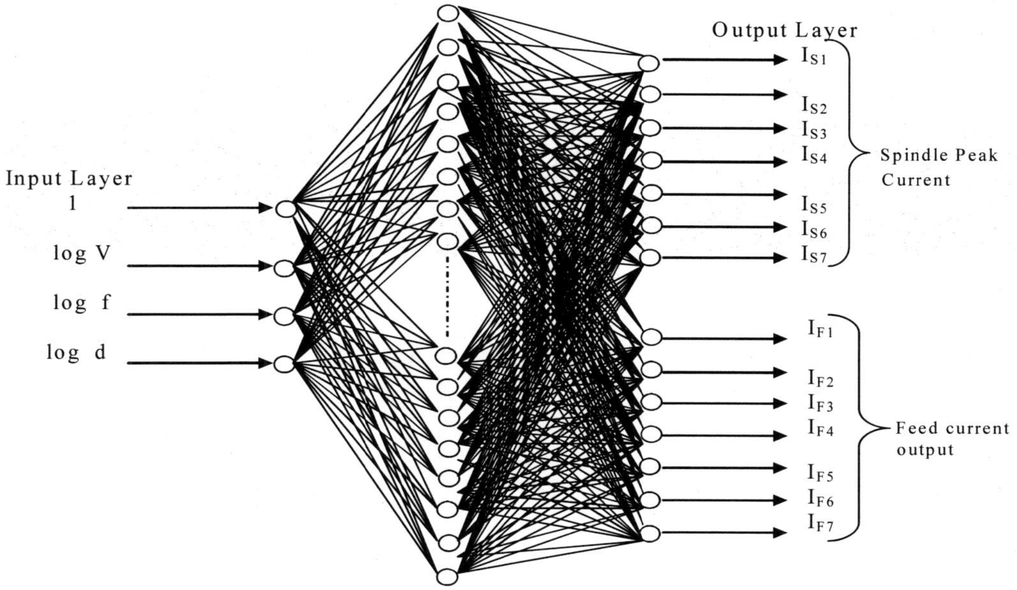

Figure 3. Spindle current signal and feed current signal with Neural Network.

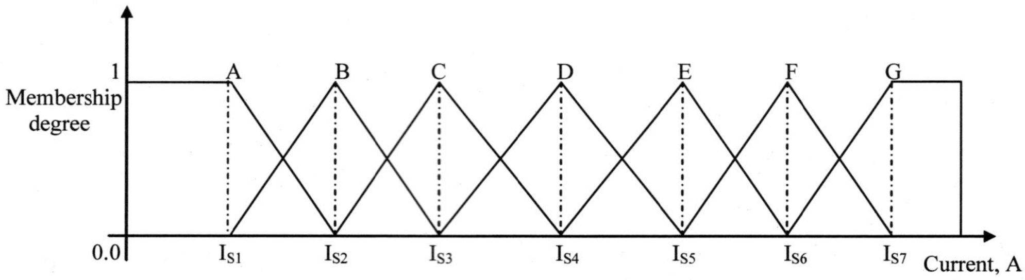

Figure 4. Fuzzy logic based curve for classify the membership degree of the tool wear.

4. Results and Discussions

AE signal generated during metal cutting mostly originates in the plastic deformation zones on the work piece and related to the associated stress relaxation, in which stress energy in the materials gets dissipated as thermal energy and micro-elastic pulses. A large number of pulses are emitted by sources randomly distributed over the plastic deformation zone. The AE signal detected from the process appears as a continuous noise. Figure 5 depicts the schematic appearance of the generated AE record. Here, 1 to 21 represents the time (s) from 0 s to 1.5 s with equal increment of 0.05 s. During experiment it was found that V_{rms} increased monotonically with increase the individual setting value of all three turning parameters such as cutting speed, feed rate and depth of cut. The parameter AE-mode measures the value of the voltage amplitude corresponding to the peak in the amplitude distribution of the digitized AE signal envelope.

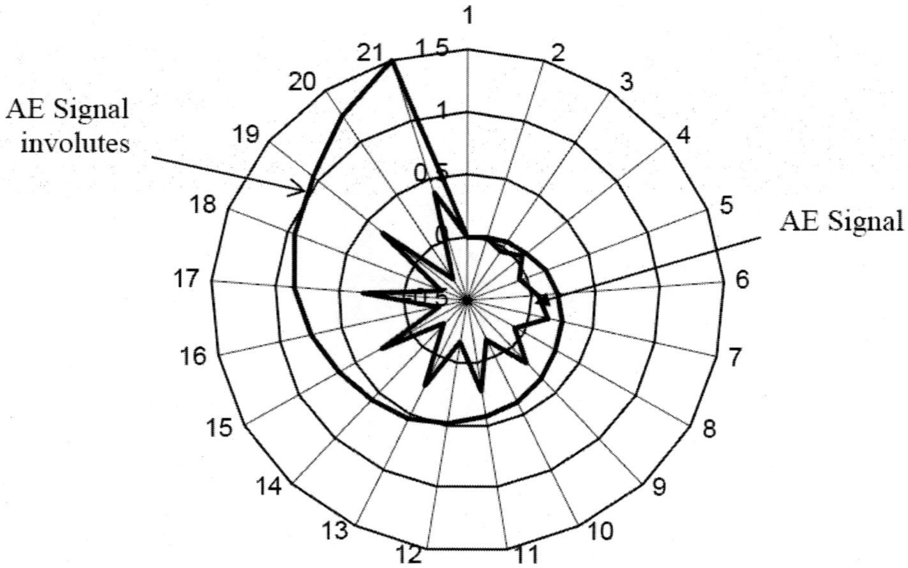

Figure 5. AE signal at cutting speed (Vc) = 100 m/min, feed rate (f) = 0.32 mm/rev and depth of cut (d) = 0.75 mm.

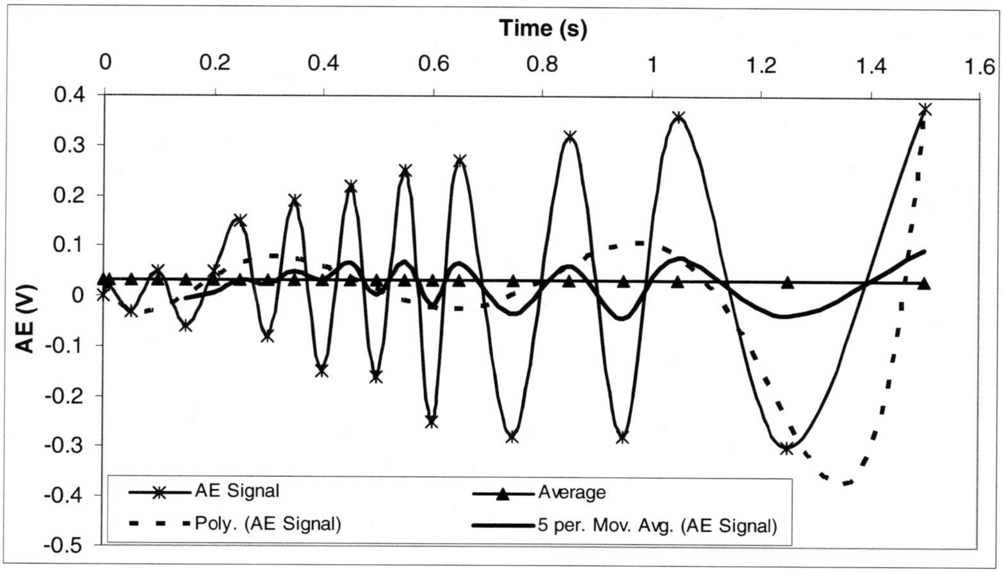

Figure 6. shows Polynomial trend and Moving Average trend of AE signal.

It is important to extract the feature of the signals during turning to ensure the reliability of the tool monitoring system. In monitoring the cutting tool flank wear AE signal monitored the complicated information during turning.

Figure 6 shows the detail AE signal generated during turning through various graphical representations such as average, polynomial trend and moving average trend of the AE signal. Polynomial equation given below represents the polynomial trend of the cutting tool wear AE signal during turning.

$$\mathbf{y} = 18.208x^6 - 73.261x^5 + 108.62x^4 - 72.722x^3 + 21.302x^2 - 2.0708x + 0.0293 \text{ ---- Eqn.5}$$

where, 'y' represents the cutting tool wear AE(V) signal, 'x' represents the time in second. Figure 6 also shows the 5 % moving average trend of the generated AE signal during turning. Average AE signal generated is slightly above the zero value and as shown in the figure 6. The behavior of the AE signal due to the appearance of the cutting tool wear can be predicted from the above polynomial relation. The developed equation 5 also can be utilized to predict the AE signal at any instant by putting the time (s) of continuous cutting. From the polynomial trend of the AE signal and mathematical relation it is clear that AE signal increases rapidly after 1.2 s of the turning operation was started. Hence, it is concluded that at the very beginning the AE signal was very small as the cutting tool was fresh and sharp edge, but after certain time (s) the AE signal increases as the cutting tool wear appears and increases with time phenomenon. During experiment it was observed that the generated AE signal increases with increase of cutting time.

According to the Taguchi method, $L_{27}(3^{13})$ orthogonal array, turning parameters and their levels (table 4) were utilized and 27 x 3 = 81 experiments were performed i.e. repeating the each set of experiment three times and taking average flank wear to obtain the real values of cutting tool flank wear. The membership degree of the cutting tool wear under different tool wear classification is estimated and fused using fuzzy logic and plotted in the graph is shown in figure-7. The figure also shows the actual and estimated flank wear width (mm) against each of the tool wear classification. From the figures 5, 6 and 7, it is clear that the above described method is suitable to estimate the cutting tool wear during turning.

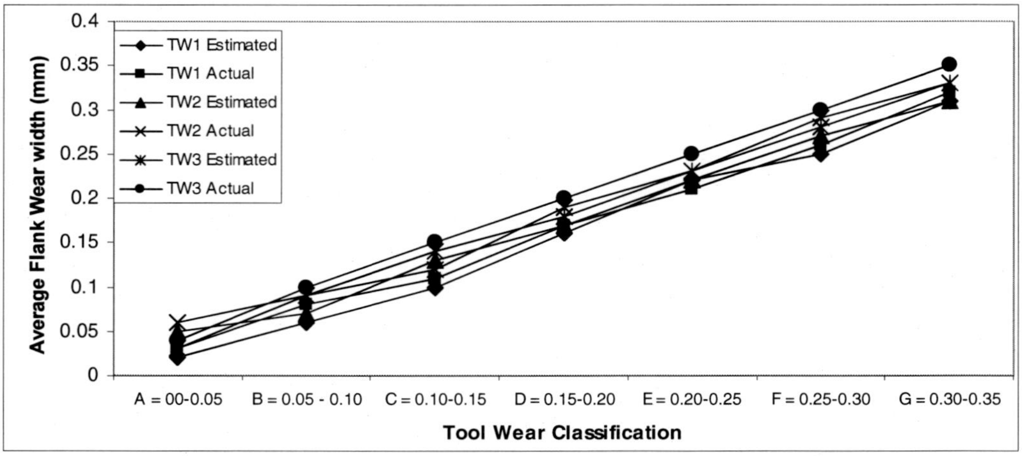

Figure 7. Cutting tool wear classification with comparison between estimated and actual wear.

CONCLUSIONS

The study on AE during turning has lead to a better understanding of the process and AE signal parameters such as V_{rms} ,I_{Si} (i = 1,2.3-------7) and I_{Fi} (i = 1,2,.....7) is to be used in tool wear monitoring. The AE signal can be successfully employed to detect the tool failure or breakage during turning of Al/SiC-MMC. AE signal levels are very sensitive to the cutting tool wear, also strongly depends on the cutting speed and depth of cut. At the beginning the AE signal was very small as the cutting tool was fresh and sharp edge, but after certain time (s) the AE signal increases as the cutting tool wear appears and increases with time phenomenon. The use of artificial intelligence technique such as FNN to offer a lot of promise, as the method work well even when exact sensor modules are not available.

REFERENCES

Tlusty,J.; Andrews,G.C (1983) "A Critical review of Sensors for Unmanned Machining" *Ann CIRP* 32, pp. 563-572.

Rao, B.C; Shin, Y.C.(1999) "A comprehensive dynamic cutting force model for chatter prediction in turning" *Int. Journal of Machine Tools and Manufacture*, vol.39,pp.1631-1654.

Chungehoo, C; Saini,D.(2000) "The total energy and the total entropy of forces signals-new parameters for monitoring oblique turning operations" *Int. Journal of Machine Tools and Manufacture,* vol.40,pp.1879-1897.

Usui,E; Hirota,A.(1978) "Analytical prediction of three dimensional cutting process part 2:chip formation and cutting force with conventional single point tool" *ASME Journal of Engineering for Industry*, vol.100(2), pp.229-235.

Usi,E; Hirota,A; and Masuko,M (1978) "Analytical prediction of three dimensional cutting process part-1: basic cutting model and energy approach" *ASME Journal of Engineering for Industry*, vol.100(2), pp.222-228.

Iwata,K.; Moriwaki,T(1977) " An application of acoustic emission to in-process sensing of tool wear" *Annals of the CIRP*; vol.26, pp.21-26.

Donfeld,D.A.(1990) "Newral network sensor fusion for tool condition monitoring" *Annals of the CIRP;* vol.39, pp.101-105.

Chaudhury,S.K; Bartarya,G.(2003) "Role of temperature and surface fenish in predicting tool wear using neural network and design of experiments" *Int. Journal of Machine Tools and Manufacture,* vol.43,pp.747-753.

Elanayar, S; Shin,Y.C (1995) "Robust tool wear estimation with radial basis function neural networks", *ASME Journal of Dynamic Systems, Measurement and Control*, vol.117, pp.459–467.

Dimla, D.E; Lister,P.M; Leighton,N, (1997) "Neural network solutions to the tool condition monitoring problem in metal cutting—a review critical review of methods", *International Journal of Machine Tools and Manufacture,* vol. 39,pp. 1219–1241.

Ghasempoor,A; Jeswiet,J; Moore,T.N, (1999) "Real time implementation of on-line tool condition monitoring in turning", *International Journal of Machine Tools and Manufacture*, vol. 39 ,pp. 1883–1902.

Li, X.P; Iynkaran,K; Nee,A.Y.C.,(1999) "A hybrid machining simulator based on predictive machining theory and neural network modeling", *Journal of Material Processing technology*, vol.89/90 , pp. 224–230.

Liu,Q; Altintas,Y, (1999) "On-line monitoring of flank wear in turning with multilayered feed-forward neural network", *International Journal of Machine Tools and Manufacture,*vol. 39, pp. 1945–1959.

Chaudhury, S.K.; Jain,V.K; and Rama Rao;Ch.V.V.(1999) "On-line monitoring of tool wear in turning using a neural network", *International Journal of Machine Tools and Manufacture*, vol. 39, pp. 489–504.

Quiza, R.; Figueira, L.; Davim, J.P. (2008) " Comparing statistical models and artificial neural networks on predicting the tool wear in hard machining D2 AISI steel" *International Journal of Advanced Manufacturing Technology*, vol.37, No.7-8, pp.641-648.

Bhattacharya, A.(1984), *Metal Cutting: Theory and Practice*, Central Book Publisher, Calcutta, India.

Haykin,S., (1994) *Neural network: A Comprehensive Foundation*, Macmillian, New York.

In: Artificial Intelligence in Manufacturing Research
Editor: J. Paulo Davim
ISBN 978-1-62618-557-9

Chapter 9

INTEGRATION OF PRODUCT DEVELOPMENT PROCESS USING STEP AND PDM

S. Q. Xie* and W. L. Chen
Department of Mechanical Engineering, University of Auckland,
20 Symonds Street, Auckland, New Zealand

ABSTRACT

In today's highly competitive global economic environment, enterprises solely replying on traditional technologies find themselves struggle to meet the demands of the international market that has become more and more dynamical and customized oriented. Integration plays an important role in enterprise strategy, and its benefits are enormous. The conventional integration technologies and methods have limitations for supporting information exchange and sharing in various stages of product development. This has led to problems such as information loss, data format incompatibility, one-way and static integration. In this chapter, an integrated platform based on the STandard for the Exchange of Product data (STEP) and Product Data Management (PDM) is proposed to overcome these problems. This platform is designed as a multi-layer reconfigurable architecture. Each level is devoted to a specific task and is designed to bring the interface with the integrated platform from different level of abstraction and functionality. Two key issues are discussed in detail in the chapter. One issue is the product modeling which is solved through a proposed Generic Product Model Data (GPMF). The GPMF consists of an EXPRESS data model (EDM), a STEP-based modeling environment, a 'five-phase' modeling method, and three EDM data exchange and sharing methods. The other key issue is process modeling, which is overcome by a proposed workflow modeling method. The workflow modeling environment is comprised of one engine, three models and one application module. Case studies are carried out to validate the proposed integrated platform, and the results show that the integrated platform is compatible, comprehensive and flexible, and is able to support product data exchange and sharing in a dynamical and bidirectional manner.

* Corresponding author, E-mail: s.xie@auckland.ac.nz

1. INTRODUCTION

In today's highly competitive global economic environment, the demand for high quality products manufactured at lower costs with shorter development times has forced most of the manufacturing industries to consider various new product design, manufacturing, and management strategies [1]. One of the strategies is to combine the strength of many companies for the development of products, especially for high value added customized products. Normally it has become very a common practice if a product is designed in a place, while manufactured and assembled in other places. This situation can be characterized by the following buzzwords: globalization, customer satisfaction, parallelization, agility, virtual enterprise, total quality management [1]. This strategy requires manufacturing companies, especially One-of-a-Kind Production (OKP) companies, to work together closely. Special critical problems for these OKP companies include high customization, 'Once' successful approach, loose or fatter production, and complicated product data and information flow [2-4]. The exchange and sharing of product and product related information has become more and more important for them and their partners. Thus, there is an urgent need for better process management and more integration within more decentralized and modular individual OKP companies as well as among companies cooperating on collaborative projects.

With the rapid advancement in information technology, computer-aided solutions such as Computer-Aided Design (CAD), Computer-Aided Process Planning (CAPP), Computer-Aided Manufacturing (CAM), Enterprise Resource Planning (ERP), have been widely used in many enterprises. These solutions or systems are, however not necessarily connected in a seamless way, thus many "islands of automation" are formed in a company that may result in communication error, rework and duplication. These "islands of automation" need to be conquered to better suit the collaborative and distributed environment. The fundamental requirements for a new OKP information system are identified as enterprise integration, open and dynamical structure, supporting cooperation and collaboration [3, 4]. It is evident that many OKP companies are struggling with integration strategy at different levels other than the underlying technology. In some cases, where integration does not exist among these computer-aided solutions, promising product technologies may come to a sudden halt against these barriers[5]. Benefits of integration to manufacturing companies has been identified[6], such as enabling a functional unit of an enterprise to communicate easily with other relevant functional units, accurate data transfer among the manufacturing plant, and/or subcontracting facilities; and faster response to required changes and increase flexibility towards introduction of new products.

To thrive in the competitive market, OKP companies must restructure their processes (by administrative, technical, or support processes), improve the way they manage these processes, and find a way to make them more integrated. Considerable research effort has been placed in this area. However, so far, the integration is still limited. This chapter proposes an integrated platform based on STEP and PDM. The platform serves as a substrate for integrating software tools and systems for OKP companies such as CAD, CAPP, CAM and ERP. The goal of the research is the creation of a technological infrastructure for a new, open integrated platform, which can adapt to continuous changes in OKP product development. A number of recent developments are discussed. They include the integration based on STEP and integration based on PDM. A prototype version of the platform has been successfully developed and demonstrated with case studies.

2. LITERATURE REVIEW

For manufacturing companies to survive and thrive in the current competitive global customer-driven market environment, technologies for supporting integration play a significant role to improve their performances. The potential benefits of the integration have attracted great research interest in this area.

2.1. Integration Based on STEP

STEP[7], formally known as ISO 10303, is an international standard designed to enable the exchange of product data between heterogeneous computer systems used throughout a product life cycle. STEP provides both broadly useful data modeling methods and data models focused on specific industrial uses. Considerable effort has been placed on STEP-based integration over the years.

Li, et al. developed a feature based parametric product modeling system, which employed a product model based on the STEP and managed by an object-oriented database[8]. This system was suitable for applications in a computer integrated manufacturing environment. Gu, et al. developed a STEP-based generic product modeling system, which was designed and implemented according to the generic resources of STEP[9]. The system can therefore be used to integrate manufacturing activities, such as process planning and inspection planning in concurrent engineering environment. Usher et al. presented a STEP-based object-oriented product model based on STEP AP 224[10]. This model was proposed for supporting CAPP analysis. Chin et al. presented a STEP-based part information model for process planning purpose[11]. Their models included a process planning information model and a production resource information model. Song et al. presented a STEP-based die and product integrated information model (DPIIM), in which integrated resources of STEP were utilized to model six EXPRESS schemas[12]. These models could support the concurrent development of stamp and die products. Zha et al. presented a product data exchange using PDES/STEP-based assembly model for the concurrent integrated design and assembly planning[13]. Shaharoun et al. utilized STEP to describe geometric data of a particular plastic product[14]. The geometrical descriptions of the product were transferred into a CAD system to assist the design and machining of a suitable mold for the plastic product. Cai et al. proposed a method to build self-defined APs for all kind of machine parts based on the STEP[15]. They implemented this method to develop two APs for presenting the geometric data model in the cone gear product for final driver of automobile driving axle system. Zhao et al. delineated an object-oriented feature-based aero-engine blade product modeling system[16]. In this modeling system, the design platform utilized STEP to standardize data modeling and to support the information transmission from design platform to analyzing system. Jasnoch et al. developed a collaborative virtual prototyping environment to integrate existing CAD systems[17]. The underlying product model of this environment was a STEP-based integrated product model.

It can be concluded that STEP-based modeling method has become the core of integration to organize product data in the standardized representation, which greatly enhances the capability of data exchanging and sharing in the integrated manufacturing environment.

2.2. Integration Based on PDM

Another integration method is based on PDM. PDM systems are designed to control product-related data and associated workflow processes. As an integration tool connecting many different areas, PDM manages product data throughout the enterprise, ensuring that the right information is available to the right person at the right time and in the right format. In this way, PDM improves communication and cooperation among diverse groups, and forms the basis for organizations to restructure their product development processes and institute initiatives such as concurrent engineering and collaborative product development. Research work has been placed on PDM-based integration during the past few years.

The product data exchange between heterogeneous CAD and PDM systems is a crucial issue for the integration of product development systems. Oh et al. introduced a UML-based mapping methodology for the product data models to exchange the product structure data between a CAD system and a PDM system[18]. Dong et al. proposed three modes of integration of CAD/PDM, the three modes are encapsulation, interface, and seamless integration and discussed the contents and data which were involved in the integration process[19]. Eynard et al. presented a UML approach for the specifications of a PDM system implementation. The chosen object-oriented approach and the used UML diagrams for the modeling and integration of product, process, and resource data was detailed for a turboprop aircraft project[20]. The Bill of Material (BOM) was put forward to solve the information integration between the CAD system and PDM system by Qiu et al.[21]. They studied the features of the information flow integration between the 2-and 3-dimensional CAD system and the PDM system, illustrated the feasibility of integration based on the BOM. Wu et al. developed a Web-based PDM system using Web services component technology for data exchange and sharing in one company and among cooperative enterprises[22]. Chan et al. proposed a concurrent control model for PDM that could cater for version management and product architecture [23]. Meng et al. built an integrated platform based on PDM system. Information integration and application integration were achieved based on process-driven. The data can be accessed each other through using the API and integration tools provided by PDM system [24].

The research work of application system integration based on PDM is another focus, especially the integration between PDM and ERP. Hou et al. proposed a methodology for the PDM/ERP/CAD integration. The UML-based methodology was applied for the product data integration model. The PDM structure model diagram and the class model diagram as well as workflow method were introduced. By workflow method, the data publican is realized by the trigger[25]. Qiao et al. presented an information conversion and integration method between PDM systems and ERP systems. The information transformation and data synchronization between the two systems were addressed by developing a conversion mechanism, with which an engineering BOM can be converted to manufacturing BOM via process planning BOM, so as to achieve information semantic unification between the two systems[26].

3. Remaining Problems

The conventional integration technologies and methods have significantly enhanced the performance of integration among various application systems. For example, geometrics models are utilized to model product geometric information, and they can support CAD systems to exchange and share product information. However, owing to the new challenges from the manufacturing environment and information technology, there are some limitations for conventional technologies:

(1) *Point-to-point or one-to-one integration.* The conventional integration technologies are normally used to model products for supporting the integration of one or two systems such as a CAD or CAPP system or special application system. They can't be directly used in the integration of systems that are employed in other stages of product development, such as computer aided manufacture or computer aided engineering (CAE). Hence, more work should be carried out to establish seamless product data exchange and sharing across the entire product development processes. For instance, the CAD based geometric product model does not define all the necessary information for the downstream manufacturing processes. The representation format of product data can impede the fluent information flow between systems, and lead to the high development costs as a result of unnecessary costly rework and redesign[27]. With the new emerging software system, the point-to-point integration has to be upgraded.

(2) *Problem of cooperation*: many complex products are usually developed by combining the strength of several manufacturing companies. Hence, data exchange and sharing between these companies should be very efficient and effective. Most companies structure their products using different modeling methods. Hence, it has become an issue for them to cooperate with each other in support of the development of a particular product. Normally, an extra data conversion process should be carried out. This is very inefficient. Sometimes, conflicts about the model structures may lead to loss of the information which cannot be converted[27]. New modeling methods have to be developed to build up the high compatibility product model.

(3) *Static and one-way integration.* Most conventional integration technologies have normally been used to integrate static product data and normally only support what we call "One-way integration". Due to lack of dynamical and bidirectional integration mechanism, product related information may not be complete, correct and up-to-date. As the product development is a very complex process involved many stages and organizations, product information is normally produced dynamically, which might be changed or modified many times according to the customer requirements during the product lifecycle. Downstream changes related to a product are also common, hence product information must be transferred bidirectionally and dynamically among different product development stages and applications. A

suitable bidirectional and dynamical integration method is required to support the cooperation and collaboration among distributed teamwork.

(4) *Limited expansibility and opening*. Conventional integration technologies are usually focused on a special issue, for example, traditional dedicated manufacturing software/hardware systems have been designed to produce specific products. When the need arose to quickly produce customized products, the systems are not up to the job and quickly became obsolete. The systems are also standalone and there are no open standard interfaces and mechanisms for them to communicate or integrate with other systems employed by partners in the global environment. This has caused a series of problems such as product data management problems, software management problems and scheduling problems[28]. Hence, there is a strong need to develop systems that are reconfigurable and extensible for the integration to keep up with the continuous changes in the global manufacturing environment and information technologies. The integration systems should be able to dynamically integrate new subsystems or tools into the environment or remove existing systems from the systems without influencing the basic.

4. An Integrated Platform Based on STEP and PDM

This section proposes an integrated platform based on STEP and PDM that aims to provide an open and extensible infrastructure for product data exchanging and sharing in the current manufacturing environment. In terms of the problems mentioned above, the research work aims mainly focused on product data integration and integration framework. The goal of the research work is to propose a generic integrated platform, which is reconfigurable and supports efficient product information exchange and sharing dynamically and bidirectionally.

4.1. Integrated Platform Architecture

An integrated platform can be characterized by a set of methods and accessible mechanisms able to support and assist in data and application integration tasks. Due to the complexity and diversity of data and application, the platform is proposed to meet the following fundamentals.

(1) Open and expandable. Information technologies used in various OKP companies are very different, with the emerging of new technology, there are always various applications and data, which are needed to be integrated. In addition, different industries may require different type information for the product development and manufacturing. The integrated platform should be open and expandable to satisfy the new emerging application integration requirements, not for some particular applications or for company current applications only.

(2) Dynamical and bidirectional integration. This is also very important for the integrated platform. As product development process normally involves activities in various stages, product related information is often transferred from one person to

another person, from one place to another place, so product information is very active, the data exchange and sharing needs to be dynamical and bidirectional.

(3) Reconfigurable and flexible. Most OKP companies are no longer geographically confined; they are run in virtual environments in which related manufacturing operations may be widely distributed geographically. The size of OKP companies also differs significantly. The integrated platform needs to be reconfigurable for various OKP companies. The function components of integrated platform must be plug-in and take-out and the system architecture should be reconfigurable and flexible.

A multi-layer, reconfigurable integrated platform architecture is proposed based on above fundamentals, as shown in Figure.1. The architecture can be described by several levels. Each level is devoted to a specific task and is designed to bring the interface with the integrated platform from different level of abstraction and functionality. The main goal of such architecture is to facilitate the application's integration task, offering different level of access and integration with the platform, providing an integrated and concurrent environment with which employees of company can share product development software tools and data dynamically, and work concurrently for the development of OKP products.

Figure 1 shows the proposed architecture that comprises five main layers; each one corresponds to a specific task.

(1) Support layer provides the necessary computer net environment that supports the system running. It includes a computer network, a Database Management System (DBMS), Operation Systems (OS), desk PCs and Servers.

(2) Database layer is the core database for the integrated platform, which stores the company's product design data, manufacturing data, industry resource data and knowledge database. Product design data can be product design models, drawings, product structure, part and other product design related information. Manufacturing data includes manufacturing process information, machining operations, cutting characteristics, tool requirements, NC codes etc. Industry resource data includes the resources such as material, machine tool, fixture database, tool which helps to guide the engineer to design or manufacture product. Knowledge database is a set of industry and company product design and manufacturing knowledge, such as the process design rules, the tool selection rules, the cutting parameter selection rules etc.

(3) Platform layer is comprised of product modeling, process modeling, enterprise application integration, which is the core and important layer. Product modeling is to provide environment and methods for modeling various types of customized product for supporting efficient data exchange and sharing. Process modeling is used to provide modeling environment and methods for modeling the product development process, such as the new product development workflow, the engineering changing workflow etc. This helps to promote the product data moving and sharing among team members and different departments.

(4) Application integration is to provide standard methods and open interfaces for integrating various applications into the platform. These applications might be used in different product development stages, such as CAD in design stage, CAPP in process planning stage, CAM in manufacturing stage.

(5) Function layer is the functional implement layer for purpose of product data management. Based on the database layer and platform layer, the function modules in this layer can read/write, present, modify, compare, and transfer all the product data stored in the database of integrated platform. For the purpose of dynamical data exchange and sharing, a workflow management module and a project management module are also provided. The organization and authorization management module is to ensure the safety of data access in the product development process. Integration interface module is a set of diverse application interfaces.

(6) Application layer is to provide industrial solution for a given company or industry, such as motorcar, tools, mould, electronics etc. Each layer of the integrated platform will be deployed flexibly according to the requirements of a particular company.

(7) As product data is required to be efficiently exchanged and shared in a product lifecycle, the integrated platform must integrate the more often used software tools in different stage of product lifecycle. These software tools can be classified by two groups, one is tools software, such as CAD, CAPP, CAM etc, the other is management software, such as ERP, CRM (Customer Relationship Management), SCM (Supply Chain Management), and MES (Manufacturing Execution System).

(8) To develop this proposed integrated platform successfully, there are two key issues need to be solved. First is about the product modeling. The goal of integrated platform is to establish seamless product data exchange and sharing across the entire product development processes and among different cooperative companies, the implementation foundation is the generic product model based on international standard, which would be accepted globally. This is the focus task and challenge for product modeling. The other key issue is about process modeling. As the integrated platform is aiming to dynamical integration distinguished from the traditional method, how to dynamically integrate the product data and related applications is also very important.

4.2. Generic Product Modeling Framework

Product modeling technology is the key and indispensable technologies for supporting efficient data exchanging and sharing in the proposed integrated platform architecture. This section introduces the generic product modeling framework (GPMF) that attempts to provide an infrastructure for modeling various types of customized products. The output of the GPMF is a set of data models defined to model a product at different stages of its development processes. Figure 2 shows the structure of the GPMF developed based on STEP. It consists of four functional components including an EXPRESS data model - EDM, a STEP-based modeling environment, a 'five-phase' modeling method, and three EDM data exchange and sharing methods.

The EXPRESS data model (EDM) is the core of the proposed GPMF. It defines a complete product data structure and uses the standardized data format. There are eleven defined EXPRESS schemas defined and STEP AP 203 included in the EDM. Each EXPRESS schema utilizes either STEP resources or STEP-based compatible resources defined to model a particular type of product information.

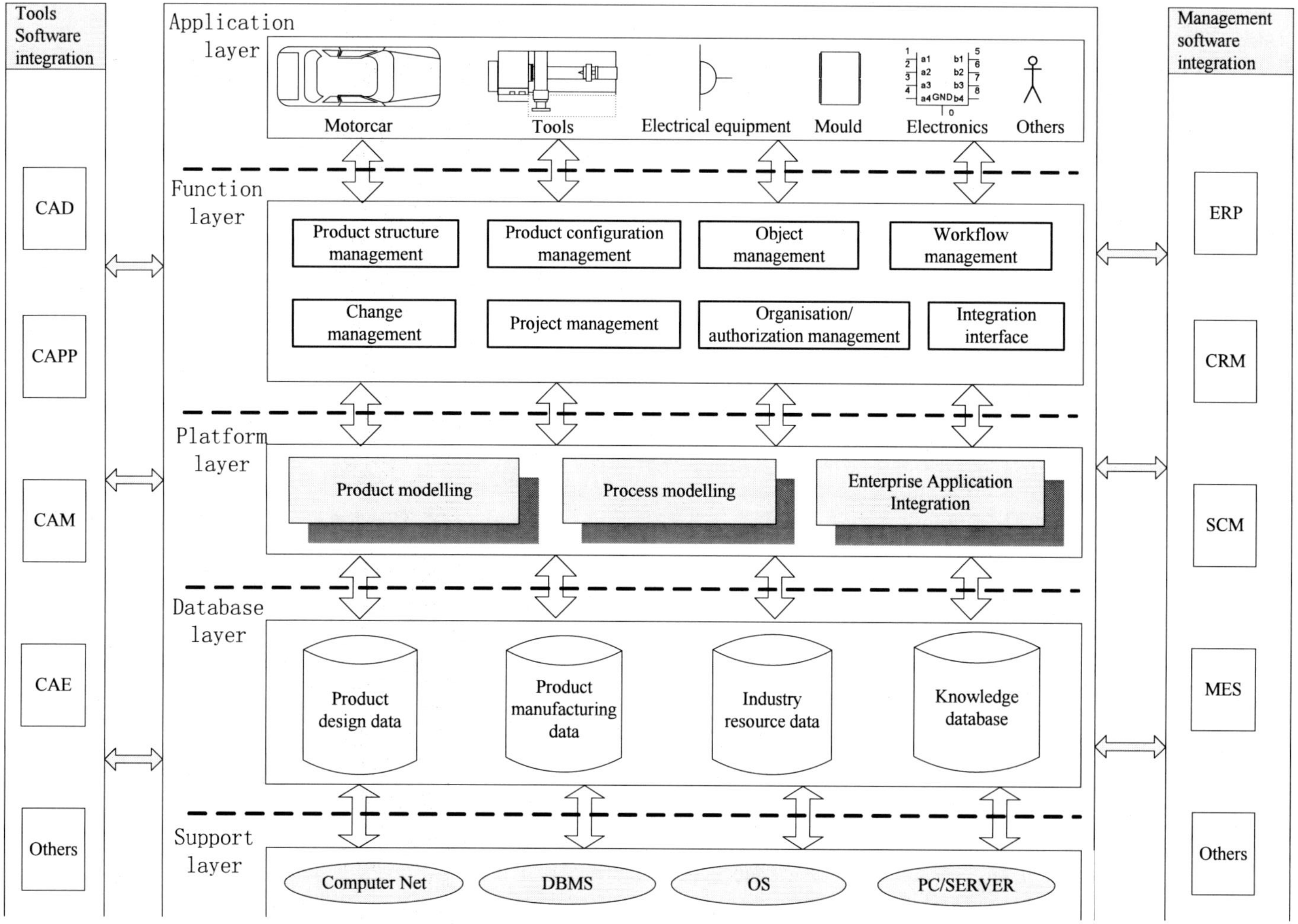

Figure 1. Integrated platform logical architecture.

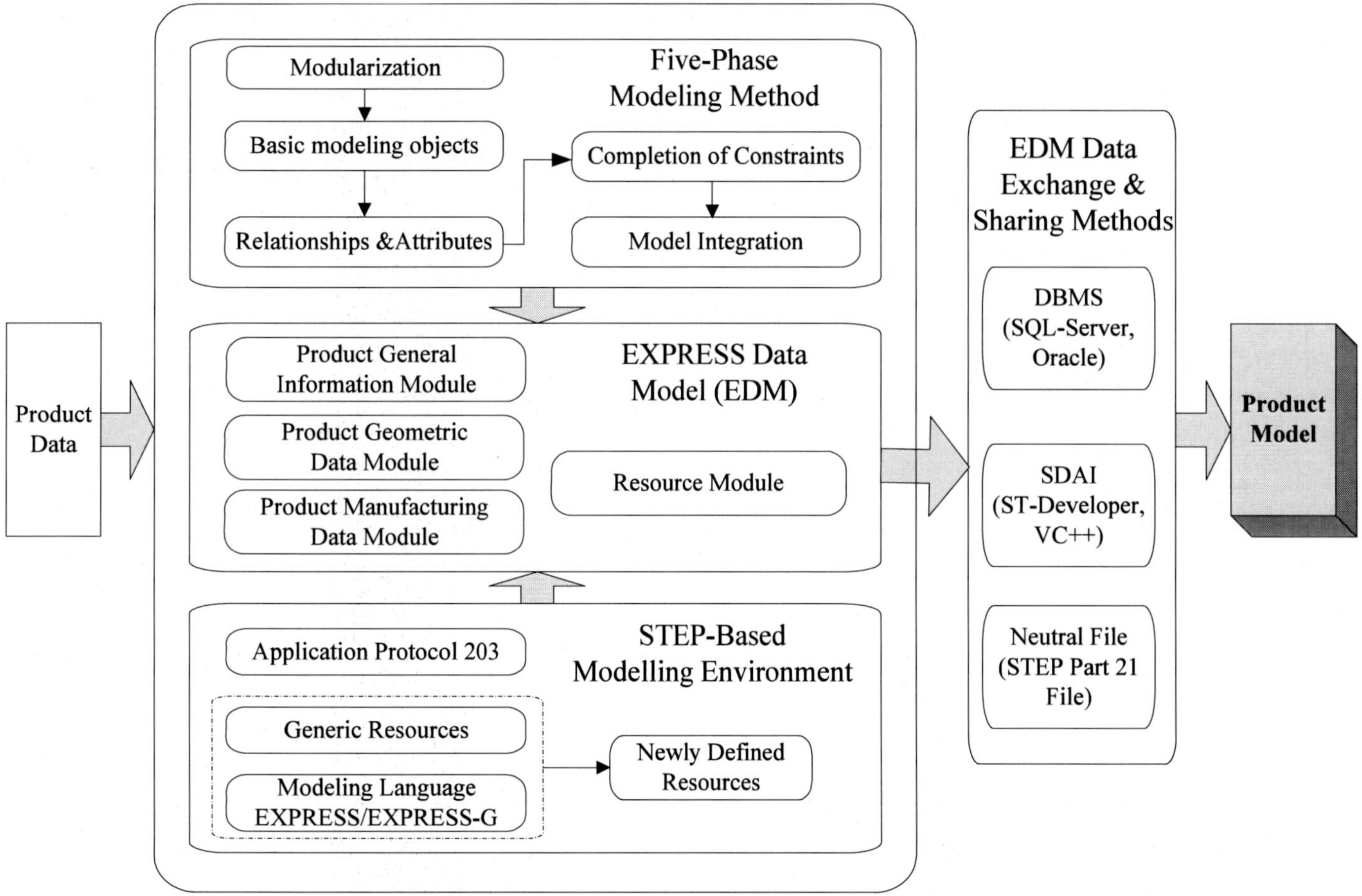

Figure 2. Structure of STEP-based GPMF.

The STEP-based modeling environment is developed for the GPMF. Within the environment, a modeling language-EXPRESS and its graphical representation method EXPRESS-G are used to model product structure. STEP generic resources are utilized to model product information defined by STEP. STEP AP 203 is used to model product geometric information, and new modeling resources are defined for modeling product information that is not covered in STEP.

The 'five-phase' modeling method is proposed to develop the EDM. The method defines a formal approach to logically organize all the tasks for building up the EDM in the modeling processes.

Three EDM data exchange and sharing methods are used in the GPMF. As shown in Figure 2, product data is exchanged and shared through either neutral physical files(STEP Part 21 file), or Standard Data Access Interface(SDAI), or Database Management Systems(DBMS). The product models defined within the GPMF are exchanged or shared using one of the methods. These three methods are easily integrated into any application software environment, which makes it easy to implement the product models defined by the GPMF in applications.

4.2.1. STEP-Based Modeling Environment

The STEP-based modeling environment in the GPMF is to facilitate data sharing and exchange through well defined STEP-compliant product data models. The EXPRESS data modules and the modeling methods are used to develop STEP-compliant data models for the integration of product development systems. A number of methods and resources are developed for establishing such an environment.

(1) Modeling Language: EXPRESS and EXPRESS-G

EXPRESS modeling language consists of language elements that can allow unambiguous data definition and specification of constraints on the data defined, and is one part of STEP defined in Part 11[29]. The resources in STEP including generic resources and APs, are normally represented as EXPRESS schema. In the GPMF, EXPRESS modeling language is utilized to develop the EDM, which represents the structure of product data. The data model in the EDM is represented by one or more schemas, which "group together the modeling objects with related meaning and purpose"[30]. The most important EXPRESS language element is the ENTITY data type, which defines the objects of interest in the domain being modeled. The ENTITY is characterized by its attributes and constraints. EXPRESS language also supports various kinds of data types, including simple types, aggregations types, and constructed types.

(2) Generic Resources and STEP AP 203

The generic resources are directly utilized to define the basic data objects in an EDM. For example, entities defined in product_definition_schema in STEP Part 41 [31] are utilized to present the product general information in the EDM; STEP Part 45[32] is used to define the EXPRESS schema that represents the data structure for material information. The generic resources are defined in STEP. The use of the generic resources enhances the compatibility of the GPMF. The STEP AP 203[33] is utilized to model product geometric information in the EDM. It is employed by different CAD systems as the data model to structure product geometric data. Thus, the product geometric data module of EDM can be integrated with CAD systems using STEP AP 203.

(3) New Modeling Resources

STEP has already defined many modeling resources to support the product modeling. However, it is still under development. There are still not sufficient resources defined to present information of different types in product development processes. Hence, two types of new modeling resources are defined and utilized in the GPMF as the supplementation of STEP-defined modeling resources.

One type of new resources is called hybrid modeling resources. They are modified from the STEP generic resources. For instance, the *document_schema* in Part 41[31] is modified to generate a data model for the document information. The method_definition_schema and process_property_schema in Part 49[34] are modified to define the data model in the EDM for the process information. New STEP-compliant modeling resources are defined by the authors to model the product data which are not covered in STEP [27].

4.2.2. 'Five-Phase' Modeling Methodology

The 'five-phase' modeling method is proposed for the development of the EDM. The method is to standardize the modeling of different types of product information. It is composed of following five phases:

(1) Phase One: Modularization

The main tasks in this phase include: (a) product modeling objective analysis, (b) classifying the product data, (c) modularizing the EDM. In the modularization phase, the EDM is divided into four modules including a product general information module, a product geometric data module and a product manufacturing data module, and a resources module that is developed to present the sharable basic modeling objects extracted from the other modules.

(2) Phase Two: Basic Modeling Objects

This phase defines the basic modeling objects and the general structure of each EDM module. The phase starts by analyzing the structure of the defined EDM modules. The fundamental elements of the modules are identified and defined as the basic modeling objects. After defining the basic modeling objects, the way to structure them is analyzed and applied in the EDM. For example, in the product manufacturing data module, product assembling information is normally defined by four basic modeling objects including an assembly product object, a product component object, a subassembly component object, and an object called connector is defined to represent the connections between these objects. These objects are named as: *assembly_product*, *part*, *subassembly*, and *connector* respectively.

(3) Phase Three: Relationships and Attributes

This phase refines and enriches the basic modeling objects defined in the second phase by adding relationships to the defined objects. The tasks are included in this phase: (a) definition of attributes and relationships between entities; (b) enhancement of the defined entities; (c) definition of new entities; and (d) correctness checking. This phase must be continued until the EDM has reached the desired level of details for representing the content of objects and their relationships with other objects.

(4) Phase Four: Constraints Modeling

This phase models the constraints of an EDM. It has the three tasks: (a) definition of constraints of objects and their relationships, (b) addition of global constraints to the model, and (c) model error- checking. Constraints modeling defines the objects and their relationships based on requirements. A complete EDM module is developed after this phase. For example, one local constraint named WR1 is defined in the *part*. This constraint specifies the range of the *level_in_assembly_hierarchy* attribute, which defines that the part must be in the second or under the second level of the assembly model tree. The error-checking task detects possible grammar errors, missing information, and conflicts of constraints.

(5) Phase Five: Model Integration

This phase integrates the modules defined after the four steps to form an EDM. The main tasks involved in this phase are to check how data is represented in each module, define the inputs and outputs the four modules, and evaluate whether EDM is complete, minimally redundant, unambiguous, and error free.

4.2.3. EDM Data Exchange and Sharing Methods

To support data exchanging and sharing, it needs to be mapped to a model that is accessible through commercial software tools. There are three EDM data exchange and sharing methods used in the GPMF. These three methods are: (1) implementation via neutral file, (2) implementation via SDAI, and (3) implementation via DBMS. Figure 3 shows the different level method which may have different implementation complexity and difficulty Level.

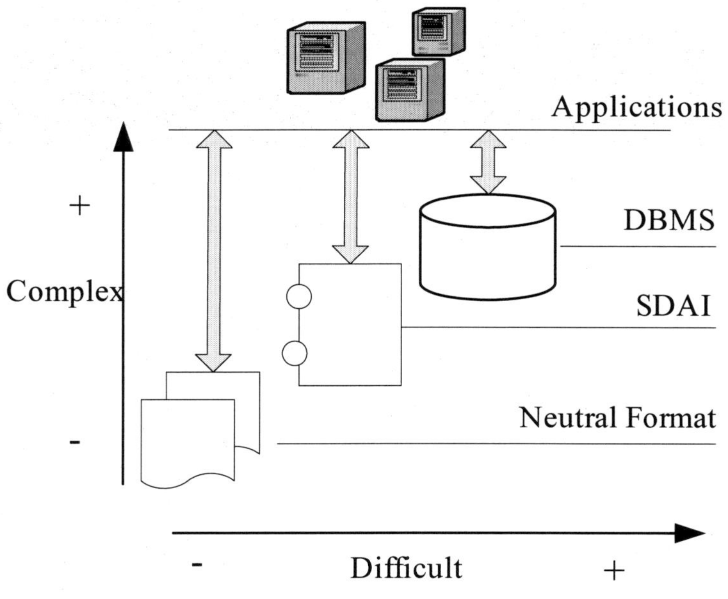

Figure 3. Data exchange and sharing methods.

EDM-defined data structures can be used to support various stages of a product development process, e.g. CAD, CAPP, CAM, etc. These systems can be integrated via the three implementation methods for EXPRESS data model. An integrated product development platform can be then developed. Product data can be made available by using API and data is presented as EDM-defined structures. Different application systems can access the product database to read and write product data. The data structure defined in the EDM should be mapped into the product database. This structure information can support the possible DBMS applications to implement this level method.

4.2.4. EXPRESS Data Model

Figure 4 shows the structure of the EDM. It consists of four modules. The product general information module, product geometric data module and product manufacturing data module are defined and based on the results of classifying product data in the first modeling phase. The resources module is developed by grouping the sharable basic modeling objects to support the development of other modules.

The product general information module represents the product data, which are not directly related with the product manufacturing, such as the product identity, product property, and the relationship between products. In this module, a *product_definition_schema* is defined to support modeling this aspect of the product data. The product geometric data module supports modeling product geometric data, such as shape information and dimension information. This module is a key to integrate different computer aided systems. In the product geometric data module, STEP AP 203 is directly utilized for representing and exchanging product 3D geometric information. The product manufacturing data module is the core of the proposed EDM, and consists of nine EXPRESSS schemas to support modeling different manufacturing data.

(1) A *supplier_informaton_schema* is defined to model the supplier information. Manufacturing companies need information about suppliers and their products to arrange the manufacturing processes.
(2) A *manufacturing_facility_schema* is defined to model the facility information. The manufacturing facilities, such as machines, tools and fixtures are directly involved in the product manufacturing processes. They are the key resources determining the product manufacturing processes and influencing the product quality.
(3) A *product_document_schema* is defined for documents aspect. The documents are defined as the industry standards and documented criteria that regulate and guide to manufacture of a product.
(4) A *bill_of_material_information_schema* is defined for representing Bill of Material (BOM) information. The BOM, which can be called part list as well, is a list of product components or resources for manufacturing or assembling this product. The BOM is utilized to determine the product cost and used as an effective tracking source during the manufacturing and assembling processes.
(5) A *material_information_schema* is defined to represent the material information. The material is the basic element for a product and it directly influences the selection of proper manufacturing facility and manufacturing processes.
(6) A *process_planning_information_schema* is defined for modeling process data. The manufacturing process information is the detailed description about the actual

manufacturing activities. It is essential to consider this aspect information in design stage, which leads to optimize the manufacturing processes.

(7) An *assembly_information_schema* is defined for these two aspects product data. Assembly product is one of the most important types of products. To develop an assembly product, the components information and the assembling method are required.

(8) An *inspection_information_schema* is defined to model inspection information. The inspection processes are essential parts of product development processes to control product quality. The inspection results can assist to indicate the problems of a product and its production processes.

(9) The *cost_information_schema* is defined for the cost information occurred in product development processes. The inclusion of cost information is critical for any engineering or manufacturing organization. Managing cost information can help the firm to increase its own competitive ability.

The resources module in Figure 4 defines basic modeling objects that are shared by the other modules. All these basic modeling objects are grouped into the *supporting_schema*. The resources in the *supporting_schema* are represented as EXPRESS entity, the constructed TYPE, and FUNTION. Through the EXPRESS schema interface, these resources in this schema are utilized by other schemas to structure an effective and efficient data model representation. The four modules of EDM are developed by applying the STEP generic resources including Part 41, Part 45, and Part 49, STEP AP 203 as well as the new defined STEP compatible modeling resources.

4.3. Process Modeling and Integration

Product development process is dynamical, which involves many sub tasks and participants. Hence, as result of the process, product data will be filled in gradually in different life cycle stages, flowed among different organizations and different software applications. The accuracy and timeliness of product data is very important during the process, which may influence the efficiency of team collaboration. Process modeling technology is the key and indispensable technology for supporting dynamical data exchanging and sharing in the proposed integrated platform architecture. This section introduces a process modeling method based on workflow modeling that attempts to provide an open infrastructure for modeling various product development process.

4.3.1. Workflow Modeling Principle

The Workflow modeling principle is described in Figure 5. The modeling environment is comprised of one engine, three models and one application module. Workflow engine is the core part of the modeling environment, which is responsible for driving workflow by interpreting workflow models, calling appropriate application system, handling related product data. Three models are organization model, workflow model and product data model. Organization model defines WHO will participate in the product development process.

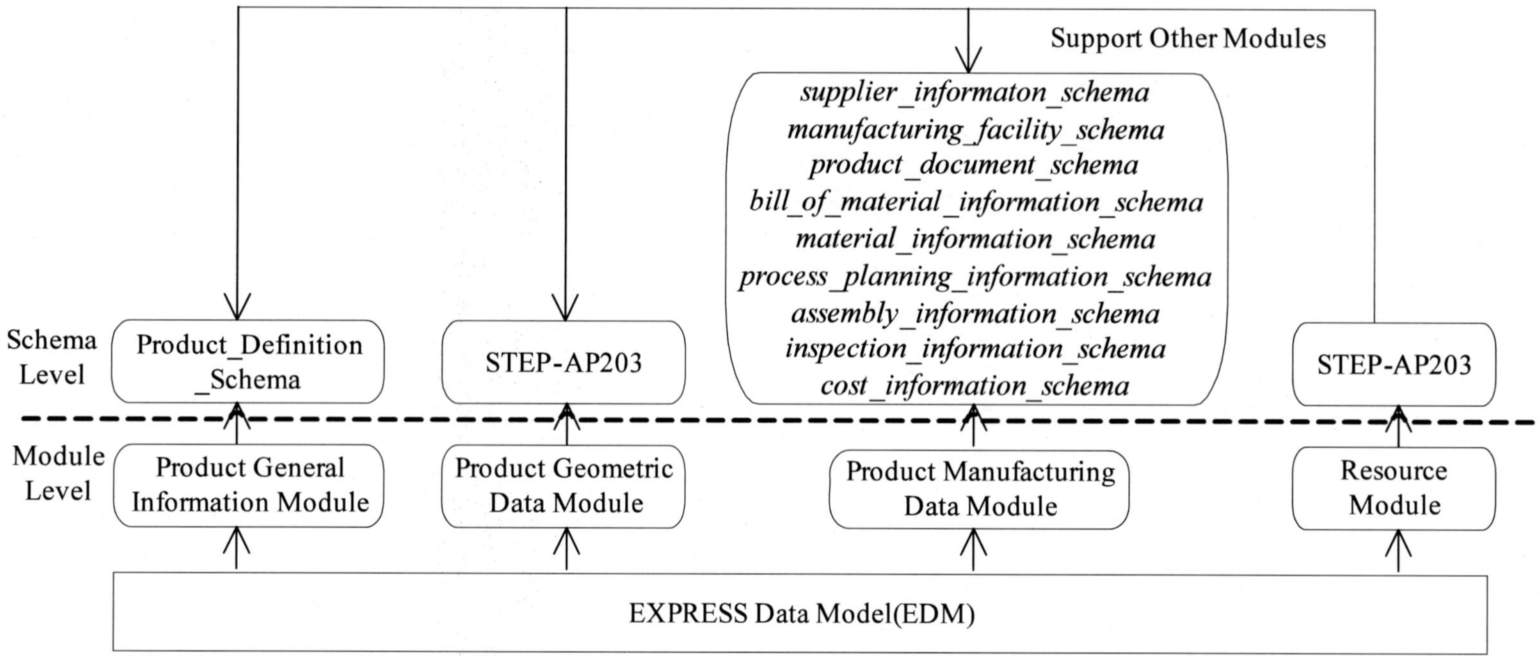

Figure 4. Structure of the EDM.

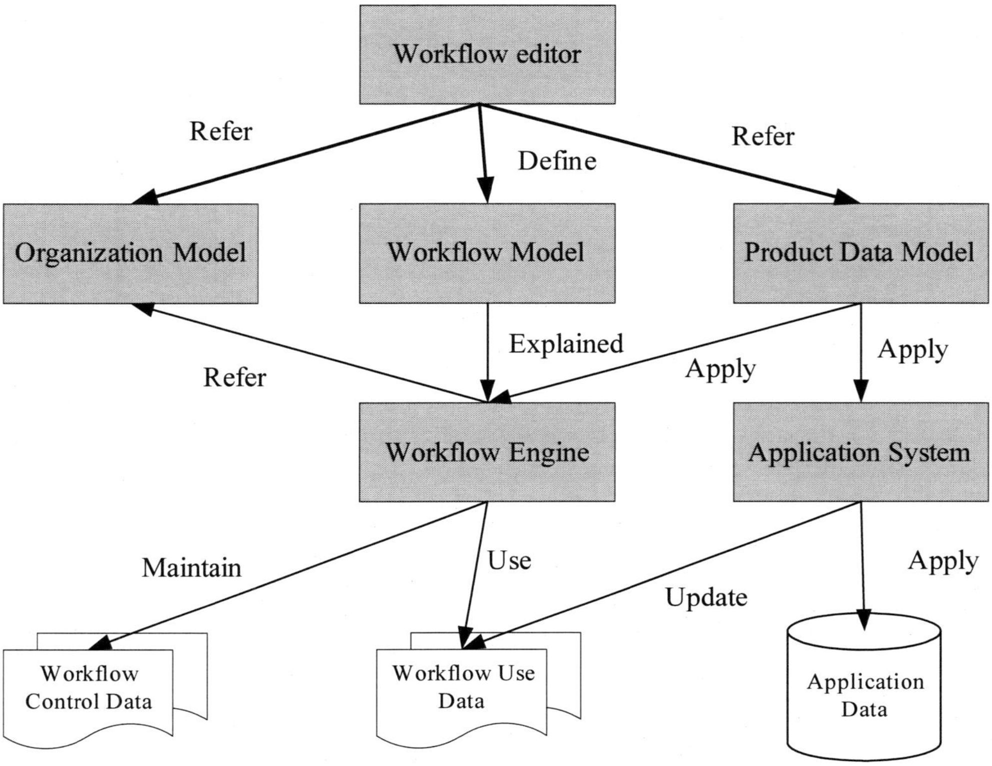

Figure 5. Workflow modeling principle figure.

This is only a logical definition, they can be totally geographically distributed organizations, teams and roles. Product data model defines WHAT to be input and output of the process; they are the data source driven by every workflow node. Application systems are registered and bound with different data type. When a type of data is dealt with, the bound application system will be called automatically to do it. For example, when a CAD file is generated by Pro/Engineering and needs to be revised, the workflow engine will help to call Pro/Engineering to open the file automatically.

Workflow model is made up of workflow templates. A workflow template is composed of a series of tasks, which may have time sequence. Task type can be classified into compound task and task step. Task step is the minimal execution task unit which can not be decomposed any more. Compound task is made up of task step or tasks. Figure 6 is a sketch map of workflow template. A, C, D are both task step, B is a compound task. The arrow in the figure indicates task or task step moving direction. These tasks or task step can be executed in serial or parallel, which are driven by the workflow engine. Task B have one input arrow, two output arrows. That means, task B receives information comes from task step A, deals with the information and judges the result according to the predefined rules, if the result meets the rules, task goes forward to task step C, if not, goes back to task step A.

Workflow editor is the core module for creating different workflow model according to the integration requirements. A workflow template is established by the workflow editor, which includes the tasks, the executive roles of these tasks and the input and output product data.

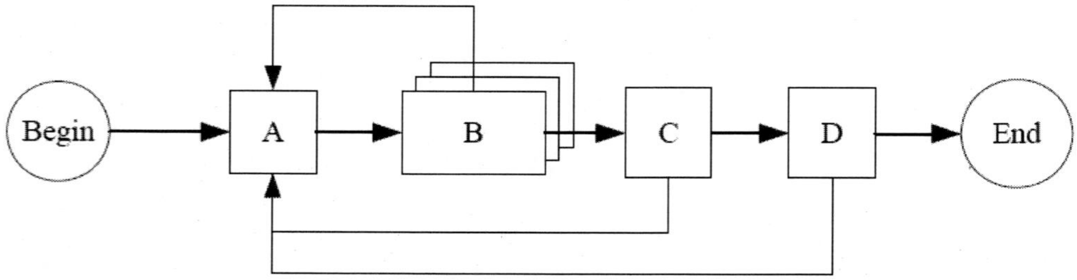

Figure 6. Workflow template.

4.3.2. Task Correlation and Status Definition

A task is divided into a series of serial or parallel sub task according to its complexity. There are different time sequence relationships among these subtasks, which are defined as task correlation. Task correlation is defined as four relationships: end-begin, begin-begin, end-end, begin-end.

- End-begin: task B can only begin when task A is ended.
- Begin-begin: task B can only begin when task A starts
- End-end: task B can only end when task A is ended
- Begin-end: task B can only end when task A starts.

As product development process is long and constantly repeated, the related task status is often changed according to inputs and outputs. Task status is shifted by workflow engine automatically, which includes seven types: initialization, wait, work, pause, termination, finish, storage. Figure 7. shows the change of task status.

- Initialization: status before a task starts.
- Wait: task is started, but cannot be changed to work status due to the restriction of task correlation. Wait status will be changed to work status automatically when task satisfies task correlation.
- Work: task has been start and in work status.
- Pause: task status is changed from work to pause. It can return to work status by activation.
- Termination: task is terminated. It cannot be restored after its termination.
- Finish: the status after task has been finished and signed.
- Storage: the status after task has been finished or been terminated. Task cannot be restored after its storage.

4.3.3. Task Step Model

Task step model is the minimal execution unit in the proposed workflow model. It is responsible for interpreting and handling the input data and requirements and exporting current task's result. A task step model is proposed as Figure 8.

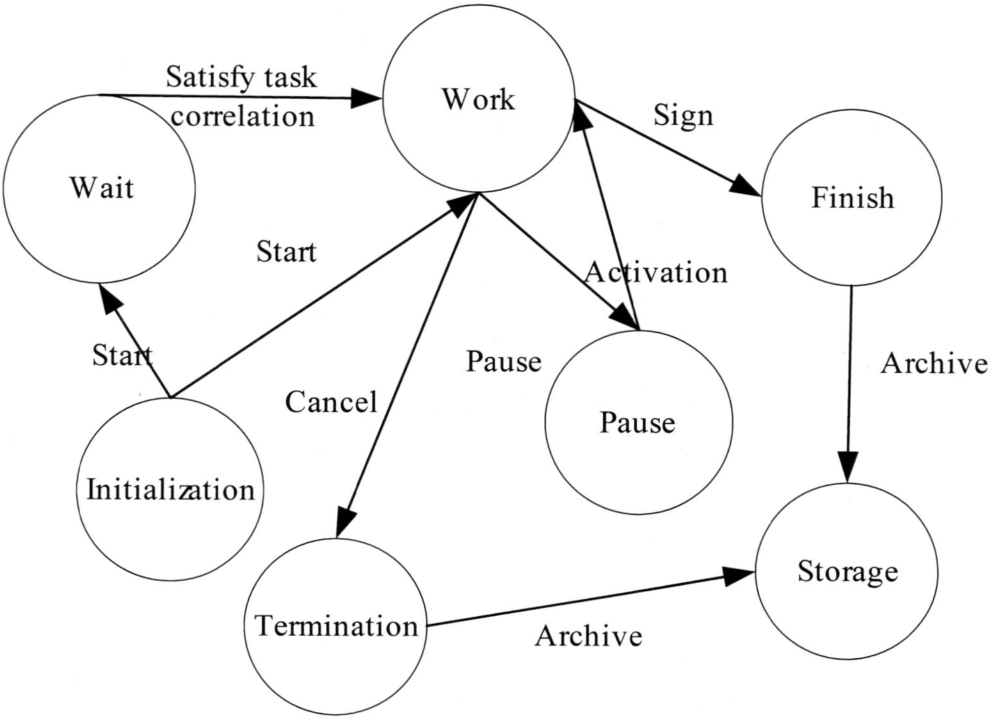

Figure 7. Task status changing in workflow.

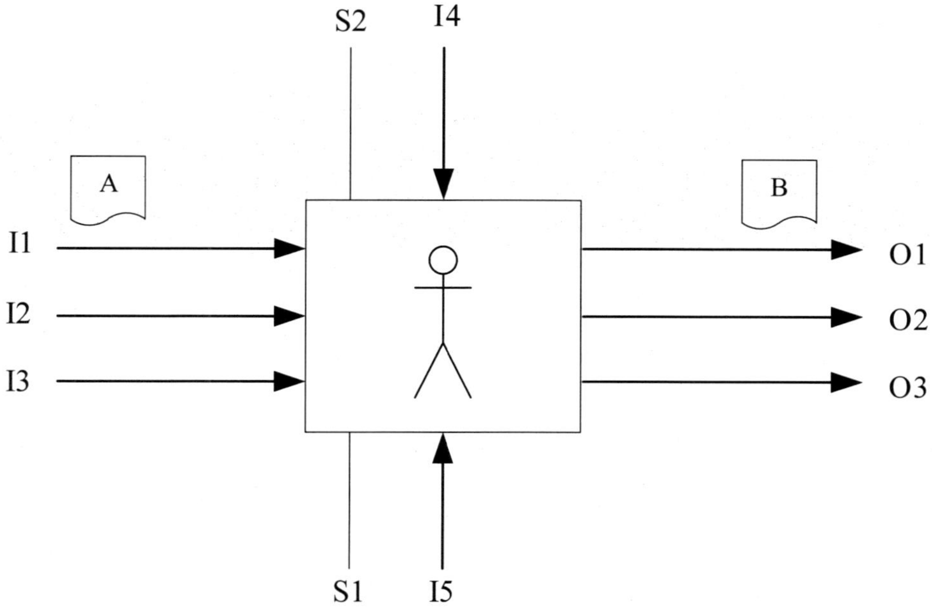

Figure 8. Task step model.

I1~I5 are the inputs of task step model, O1~O3 are the outputs of task step. S1~S2 are its own attributes. Their definitions are as following:

- I1: the input product data or files that are needed to be dealt with by the task step.
- I2: the time requirement for the task step. It defines the expectative start time and end time.
- I3: the people who will execute the task step. The people can be one or more real persons; it also can be an organization, not a real person.
- I4: authorization definition. This shows what kind of authorization to the input data and files the people may have in this task step. Normally, the authorization activities may include browsing, modifying only, marking/modifying and marking only.
- I5: signature mode. This mode defines in what situation can pass the current task step. There are four signature modes: anyone pass, all pass, half pass and two third pass.
- O1: the output product data or data files of task step.
- O2: signature opinion. After finishing current task step, the people must give opinion about current task step: agree, disagree or abstain.
- O3: the next task step or compound task after current task is finished.
- S1: task step status. The seven types of status are introduced in Section 4.3.2.
- S2: task step correlation with other task step. The four correlation relationships are introduced in Section 4.3.2.

4.4. Integration Scenario

Figure 9 shows the tasks involved in the integration of tools throughout product development life cycle, from design to fabrication and assembly. STEP and workflow are used within the cycle. The design phase results in STEP data and includes the geometry of the part, which is defined in AP203 or AP214. The data file is then transferred to macro process planning which is a high-level process plan for a machined part. The process plan information for both numerical controlled and manually operated applications, and associated product definition data is defined in AP 240. The macro process planning defines the sequence of machine process, each of them identifies the machine tool, machine setup, clamping positions, a list of machine operations, and a list of machining features that are eligible to be machined per process. Micro process planning comes after macro process planning, which is closely related to a CNC machine. This is done by AP238[35] or ISO 14649. Micro process planning defines the process information for a specific class of machine tools, such as turning, milling, drilling, reaming or tapping. The feature sets are used during the computer-aided manufacturing phase. Based on this, machine independent STEP-NC files are generated that are executed by CNC machine tools and results in the machined parts. The manufactured part is transferred to assembly line, and then the end product will be produced.

The figure helps to understand the activities in a whole product development chain. The organizations can be in one company or geographically distributed companies. The product development activities are driven by workflows automatically with related product data.

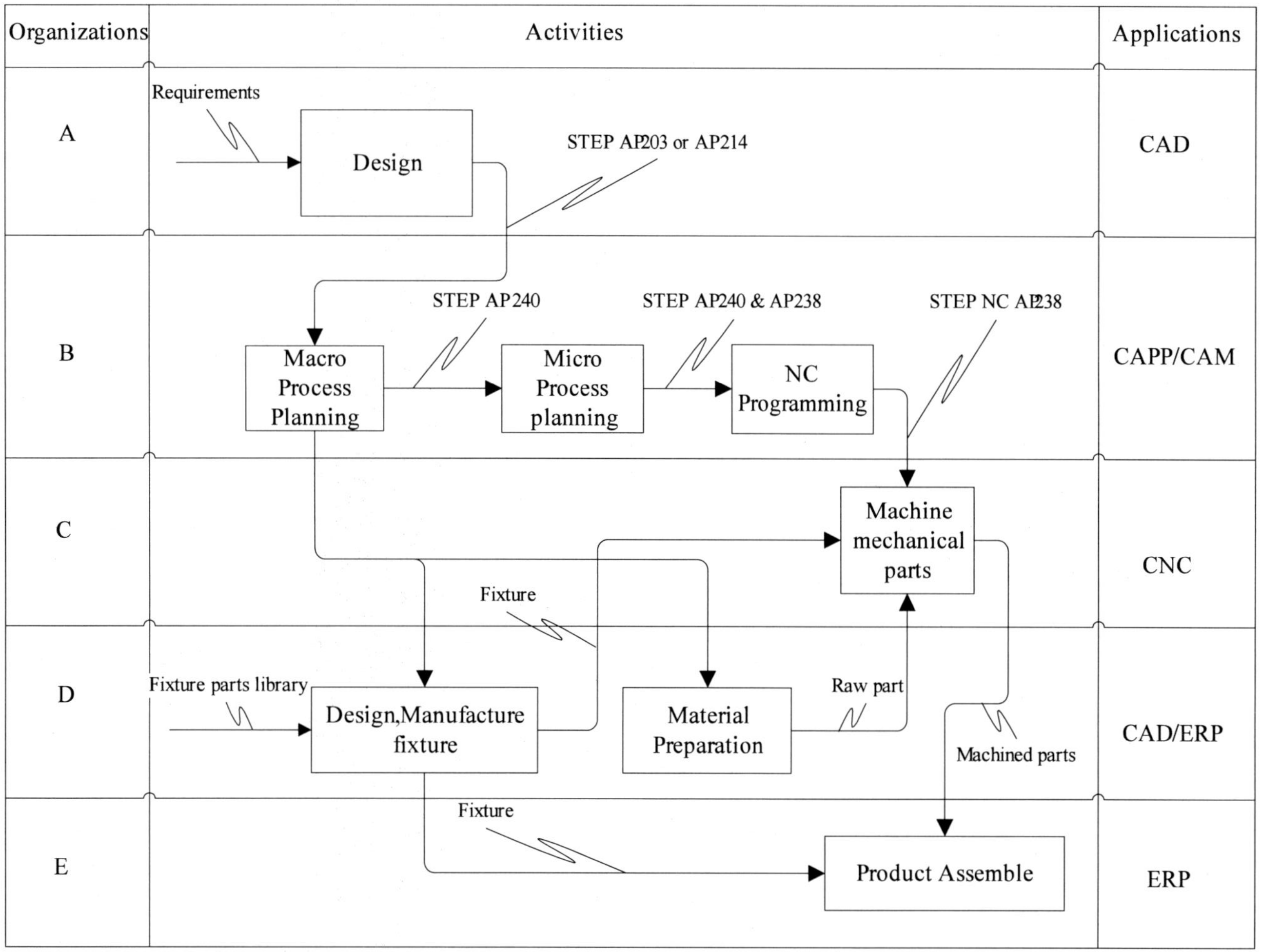

Figure 9. The integration scenario in product development life cycle.

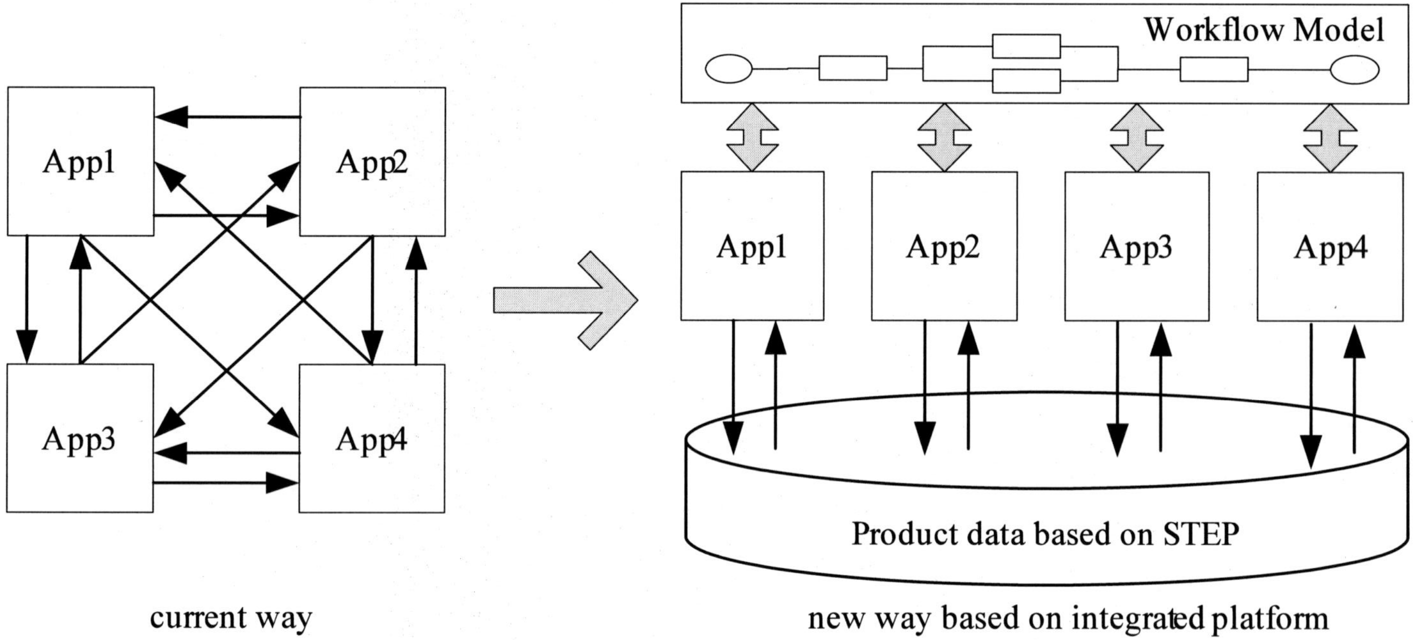

Figure 10 The different way of action in product development life cycle.

In different phases, different applications may be used in product development, such as CAD used in design phase, CAPP/CAM used in process planning phase, ERP used in manufacturing phase. This accelerates product designing and manufacturing as a whole.

The difference between current way and integration way is shown in Figure 10. As in current way, the integration between applications such as App1~4 is one-to-one, there are no close relationships among them. Product data exchange and sharing interface may have to be customized to a special application and the data flow is manual. While using the new way based on integrated platform, the data exchange and sharing is via database based on STEP, every application is interoperable with the common product database without one-to-one. The action of exchange and sharing is driven by workflow engine automatically. This improves greatly the efficiency and quality during product development process.

5. Case Study

A prototype system names Teamwork Solutions has been developed to validate and demonstrate the feasibility and the compatibility of the proposed integrated platform. Main software tools used to develop the system include: Microsoft Visual C++ 2005 which is employed to develop framework and functional modules; ST-Developer 12, which translates data between STEP information models and integrated platform; Microsoft SQL Server 2005 which is utilized to construct the system's repositories. One interface of the prototype system is shown in Figure 11.

Figure 11. Main control center interface of prototype system.

To illustrate the related functionality of the prototype system, the following main tests will be carried out:

(1) Create a product development workflow and start it.
(2) Design a product using Pro/Engineering tool.
(3) Product information integrated into system's database.
(4) Proofread the designed product.
(5) Return to design stage and revise the product.

The case studies are focused on the information and activities integration only using partial stage of product development life cycle; the other product development activities are very similar in principle.

(1) Scenario 1: Workflow modeling

An authorized person, e.g. product manager, can create company's own special new product development workflow by using workflow modeling module of system. The example of workflow created by prototype system is shown in Figure 12. A new designed product will go through several phases which are showed as task icons, each phase stands for a special task for product development, such as part designing, proofreading, verifying, process planning, standardization and confirmation. The arrow indicates the flow direction of product data and task. Every task node may have different signature options and authorization to the input data which can be defined through user interface shown in Figure 13.

After modeling the workflow, the product manager can start the workflow, and specify the executor of each phase. The first task is to design a product. When the executor of this task logins system, he will receive the task information.

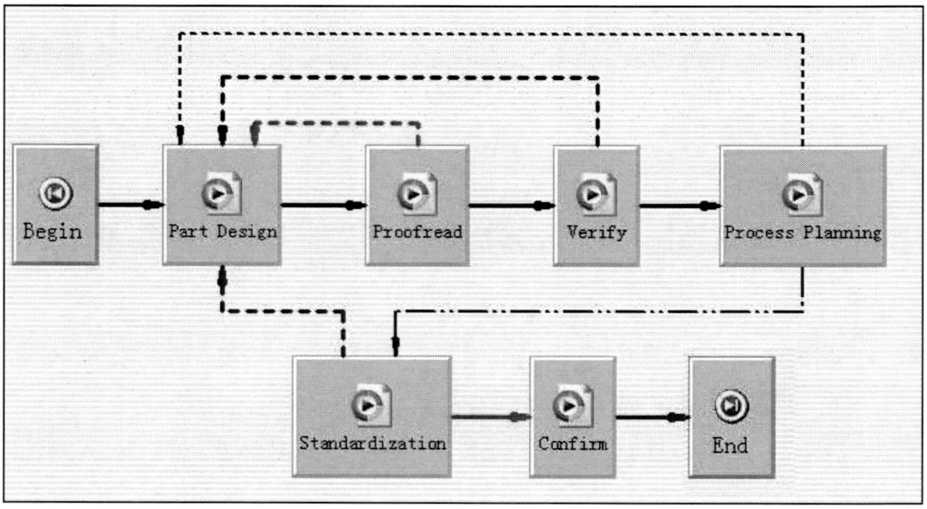

Figure 12. Workflow modeling example.

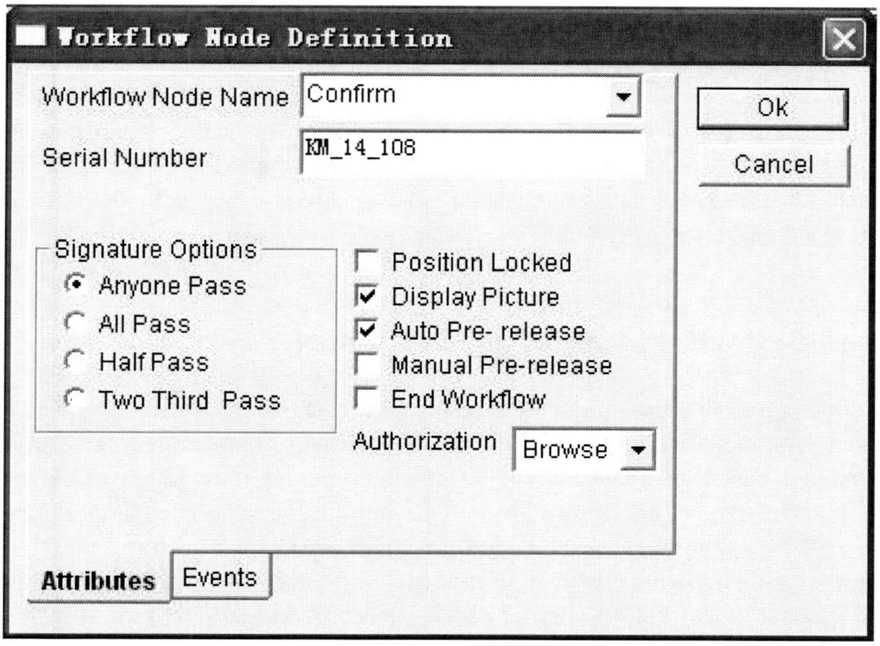

Figure 13. Workflow task node definition.

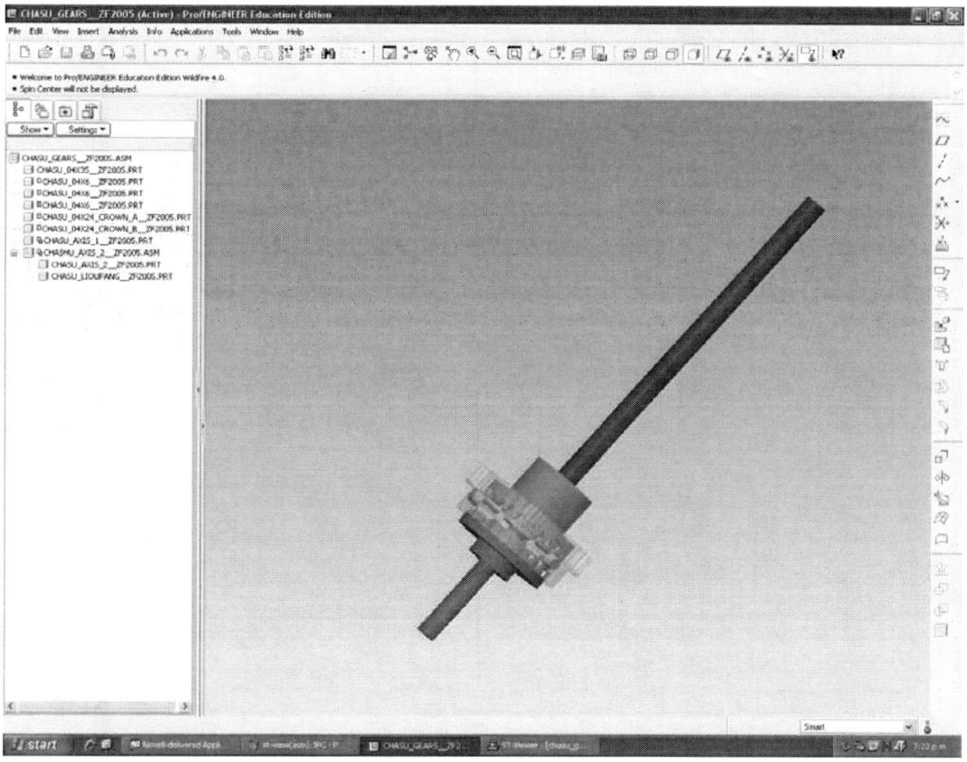

Figure 14. Transmission shaft product model designed in Pro/Engineering.

(2) Scenario 2: Part Design

The aim of this phase is to design a product using a CAD tool. The CAD tool can be registered in the prototype system in advance, which will be called when current task is executed. Figure 14 shows a transmission shaft product designed by Pro/Engineering. This product includes nine parts; its hierarchy relational structure is shown at the left side of the figure, which has two levels.

(3) Scenario 3: Product Modeling and Integration

After finishing design the product, the result will be stored in system's database in order to be exchanged and shared by other applications and other organizations. The product data will be modeled based on above mentioned methods before it is stored in database. The geometric product data of the transmission shaft product are modeled into a Part 21 file through an AP203 output interface defined in Pro/Engineering.

Figure 15 shows the windows display results of the transmission shaft when this Part 21 file format product model is loaded in ST-Viewer in the ST-Developer toolkit. It is apparent that the geometric product data in the product model can be read by systems to regenerate the 3D geometric models which has STEP AP 203 interface to load/output product geometric data. The screenshots of the transmission shaft product model in both Pro/Engineering and ST-Viewer are the same, which means that the geometric product data of the product model can be exchanged and shared by the above two application systems.

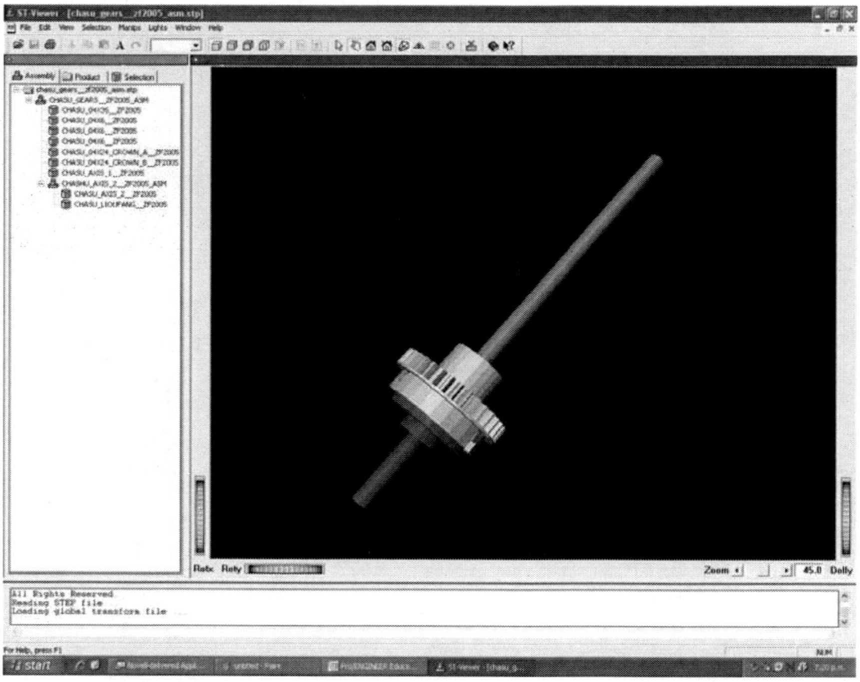

Figure 15. Transmission shaft product model loaded in ST-Viewer.

Part 21 file is the important output of design phase, it is the foundation of further product data exchange and sharing in different product development stages. The Part 21 file of transmission shaft is shown in Figure 16.

```
ISO-10303-21;
HEADER;
......
FILE_SCHEMA(('CONFIG_CONTROL_DESIGN'));
ENDSEC;
DATA;
#2=DIRECTION('',(0.E0,0.E0,1.E0));
#3=VECTOR('',#2,2.E0);
#4=CARTESIAN_POINT('',(7.394200706531E0,2.929093913363E-1,-2.E0));
#5=LINE('',#4,#3);
#6=DIRECTION('',(0.E0,0.E0,-1.E0));
#7=VECTOR('',#6,2.E0);
#8=CARTESIAN_POINT('',(6.577848345501E0,0.E0,0.E0));
#9=LINE('',#8,#7);
#10=CARTESIAN_POINT('',(6.577848345501E0,0.E0,0.E0));
#11=CARTESIAN_POINT('',(6.621001042825E0,3.472681210621E-3,0.E0));
......
ENDSEC;
END-ISO-10303-21;
```

Figure 16. Part 21 file for geometric data of transmission shaft product.

This Part 21 file can be imported by other application systems to achieve the product geometric information. The prototype system has proposed a special interface to read directly Part 21 file based on STEP AP 203. The developed user interface is shown in Figure 17.

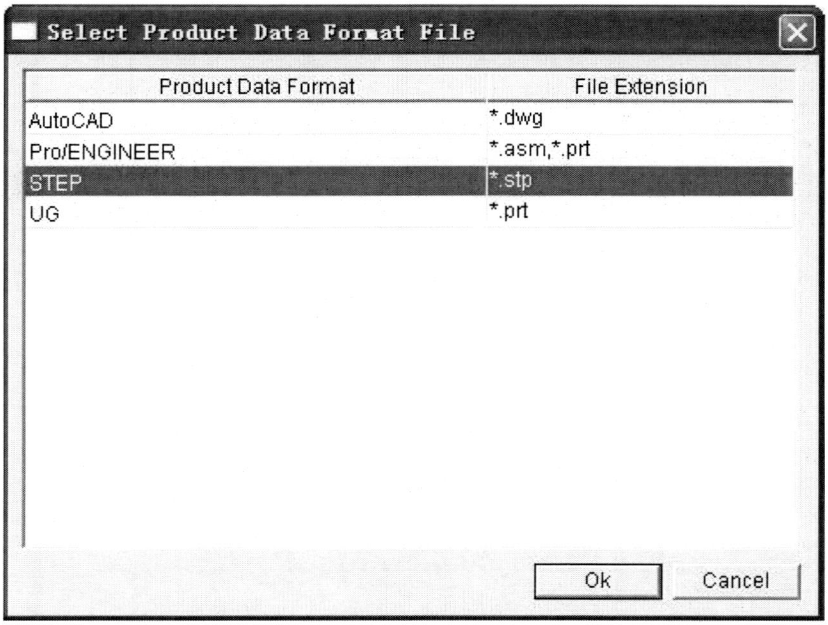

Figure 17. STEP file read interface.

After reading into the Part 21 file, the system deals with the data and stores them into its own database. Through another view, the transmission shaft product structure and 3D model can be browsed, as shown in Figure 18. It can be found that screenshots of the transmission shaft product 3D model and product structure in both Pro/Engineering, ST-Viewer and prototype system are the same, which further verifies that the product data model can be exchanged and shared among application systems when they are based on STEP.

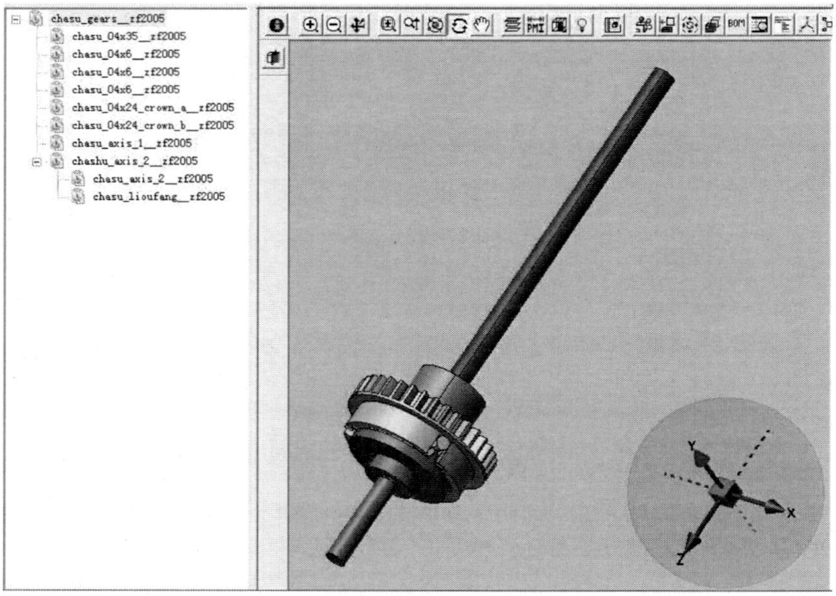

Figure 18. Browse product 3D model in system.

(4) Scenario 4: Proofread the Designed Product

As the designer finishes his task and submit the result of his task, the workflow will automatically move to the next predefined task: proofread. The executor maybe another person. He will receive the task when he logins in the system. As the design result has been stored in the database and flow automatically to current task executor, the executor can then browse the 3D model, check the structure of product etc as he likes under the given authorization. He may then give his opinion about the designed result: agree, disagree or abstain, attached with his reason and suggestion. In this example, we assume the given opinion is “disagree”, so the workflow has to be back to the front phase: part design. The interface of this phase is shown in Figure 19.

(5) Scenario 5: Return to Design and Revise the Product

When the design result is returned back, the product designer has to revise the product according to proofreader’s suggestions. This example illustrates the bidirectional integration between CAD and prototype system. The prototype system can import and access product data from CAD system, which is called "positive integration". The CAD system can also access database of the prototype system, which is called "reverse integration". When a designer logins the system, he will receive his task of revising his former designed product. Driven by the given workflow model of the prototype system, Pro/Engineering may be called and the former designed product is opened, see Figure 20.

The screenshots shows the same 3D product model and the same product structure as before without data loss. The transmission shaft product model data is transferred to Pro/Engineering positively and automatically by prototype system. The change is found at the menu of Pro/Engineering, which is added a new menu named "PDM". This menu is made up of several communication functions, which can be used to access database of prototype system under the system's authorization. These functions include "connect PDM", "disconnect PDM", "Check in", "Object information", etc. With the help of reverse integration, designer can search, compare, copy or edit product database, execute own task under CAD environment. Due to the interface of reverse integration is developed based on COM/DCOM/COBRA standard, it can also be used to integrate with other applications such as CAPP, CAM etc.

When designer has finished revising the product, he can check in his product file and submit his current task. Then the task will go forward to the next driven automatically by workflow engine until they are finished. All the product data will be transferred dynamically and bidirectionally among tasks and all will be stored in system's database together with its related process.

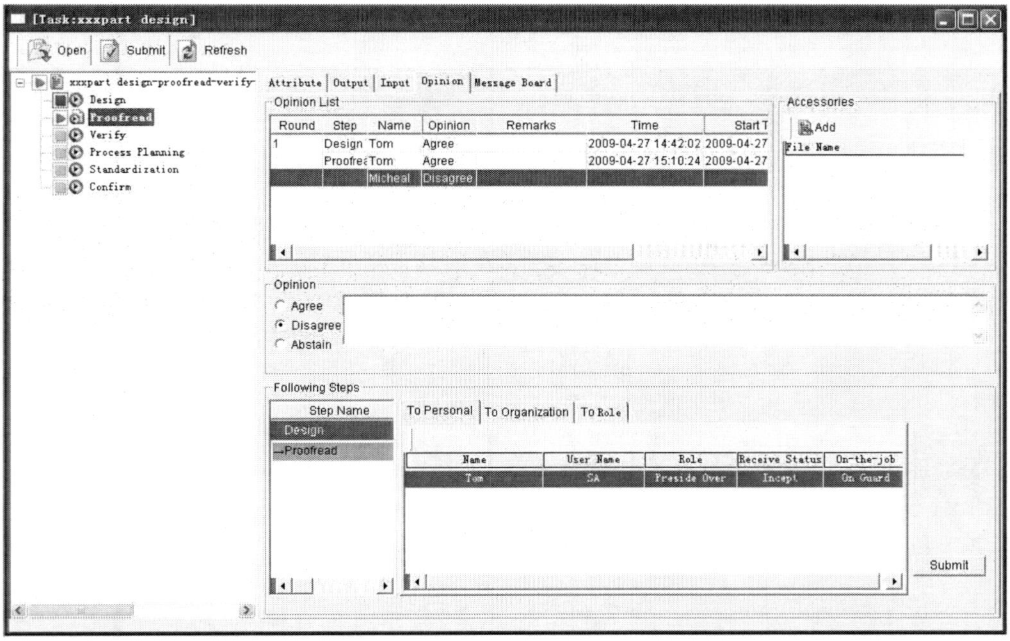

Figure 19. Interface of task step-proofread.

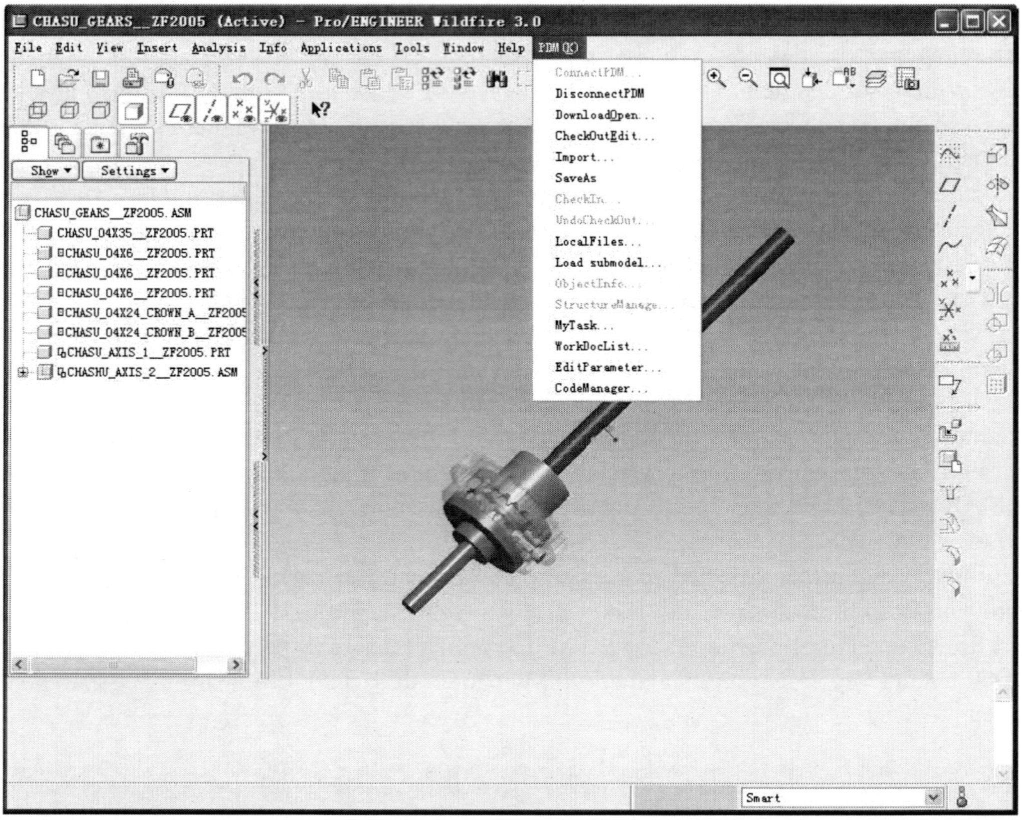

Figure 20. Reverse integration in Pro/Engineering.

CONCLUSION

In the advent of information technology and in the globalization of society, integration has become a vital issue for businesses to grow and improve efficiency. The need to exchange semantically rich information has also grown. In this chapter, an integrated platform based on STEP and PDM is proposed to integrate tools and systems in order to achieve high information exchange and sharing performance improvements for OKP companies. Two key issues are discussed in detail in the chapter. One is the product modeling which is solved through the proposed GPMF; the other is process modeling which is achieved by the workflow modeling. Case studies have been carried out to validate the proposed integrated platform. From this research work, the following conclusions are drawn:

(1) The proposed integrated platform is compatible for integration of different systems and software tools. This has been validated in our case studies.
(2) The proposed modeling methods enable the integration of various application systems bidirectional through common product database and the proposed integrated platform.
(3) The product data flow can be driven forward or back by the proposed workflow model dynamically and automatically.

(4) The dynamical integration capability of the proposed modeling scheme eliminates the need for human interpretation, errors and rework in product development life cycle.

The developed prototype system illustrates that the basic mechanisms of the integrated platform are feasible; however, future research work is still required to further improve the proposed platform.

(1) As information technology develops very quickly, new systems and tools will be developed to support product development processes, further research work is required to improve the proposed integrated platform in order to integrate these new tools and systems.
(2) Develop new product modeling methods for the integrated platform. An implementation method needs to be developed to map between extensible mark-up language (XML) and EXPRESS, which is defined in STEP Part 28[36].
(3) As STEP itself is still at its development stage, further research work should keep up with the new pace of STEP to achieve better performance, such as STEP-NC etc.

REFERENCES

[1] F.B.Vernadat, *Enterprise modeling and integration : principles and applications*. 1996, London ; New York :: Chapman and Hall.
[2] Tu, Y., Xie SQ,Zhou ZD. An information modeling framework to concurrent product design and manufacturing. in *Proceedings of IAM'2001*. 2001. American University in Dubai, U.A.E.
[3] Tu, Y.L., S.Q. Xie, and J.J. Kam, Rapid one-of-a-kind production. *International Journal of Advanced Manufacturing Technology*, 2006. 29(5): p. 499-510.
[4] Xie, S.Q. and Y.L. Tu, Rapid one-of-a-kind product development. *International Journal of Advanced Manufacturing Technology*, 2006. 27(5-6): p. 421-430.
[5] Xu, X., *Integrating advanced computer-aided design, manufacturing, and numerical control: principles and implementations*. 2009, New York: Hershey.
[6] Nagalingam, S.V. and G.C.I. Lin, CIM-still the solution for manufacturing industry. *Robotics and Computer-Integrated Manufacturing*, 2008. 24(3): p. 332-344.
[7] J.Kemmere, S., *STEP:The Grand Experience. NIST Special Publication 939*. 1999, Washington: U.S.Government Printing Office.
[8] Li, H., et al. Feature-based, parametric modeling system for CAD/CAPP/CAM integrated system. in *Proceedings of the IEEE International Conference on Industrial Technology*. 1996.
[9] Gu, P. and K. Chan, *Product modelling using STEP. Computer-Aided Design*, 1995. 27(3): p. 163-179.
[10] Usher, J.M., A STEP-based object-oriented product model for process planning. *Computers and Industrial Engineering*, 1996. 31(1-2): p. 185-188.

[11] Chin, K.S., Y. Zhao, and C.K. Mok, STEP-based multiview integrated product modelling for concurrent engineering. *International Journal of Advanced Manufacturing Technology*, 2002. 20(12): p. 896-906.

[12] Song, Y., X. Chu, and F. Cai, Real-time concurrent product and process design system for mechanical parts. *High Technology Letters*, 1999. 5(1): p. 74-80.

[13] Zha, X.F. and H. Du, A PDES/STEP-based model and system for concurrent integrated design and assembly planning. *CAD Computer Aided Design*, 2002. 34(14): p. 1087-1110.

[14] Shaharoun, A.M., J. Ab Razak, and M.R. Alam, A STEP-based geometrical representation as part of product data model of a plastics part. *Journal of Materials Processing Technology,* 1998. 76(1-3): p. 115-119.

[15] Cai, C.T., et al., Design method of application protocol of the machine parts based on STEP. Jisuanji Jicheng Zhizao Xitong/Computer Integrated Manufacturing Systems, CIMS, 2002. 8(11): p. 892-895.

[16] Zhao, W. and M. Ma, Feature modeling for aeroengine blades according to STEP. Beijing Hangkong Hangtian Daxue Xuebao/Journal of Beijing University of *Aeronautics and Astronautics*, 1999. 25(5): p. 535-538.

[17] Jasnoch, U. and S. Haas, A collaborative environment based on distributed object-oriented databases. *Computers in Industry*, 1996. 29(1-2): p. 51-61.

[18] Oh, Y., S.-h. Han, and H. Suh, Mapping product structures between CAD and PDM systems using UML. *Computer-Aided Design*, 2001. 33(7): p. 521-529.

[19] Dong, J.H., et al., Research on CAD/PDM integration. Jixie Kexue Yu *Jishu/Mechanical Science and Technology*, 2001. 20(2): p. 288-290.

[20] Eynard, B., et al., UML based specifications of PDM product structure and workflow. *Computers in Industry*, 2004. 55(3): p. 301-316.

[21] Qiu, J.X., X.W. Wang, and X.C. Fan, Information integration between computer aided design system and product data management system in collaborative product design. Jilin Daxue Xuebao (Gongxueban)/Journal of Jilin University (Engineering and Technology Edition), 2005. 35(5): p. 505-510.

[22] Wu, J.W., et al., Management strategy of product data in distributed collaborative design environment. Zhejiang Daxue Xuebao (Gongxue Ban)/Journal of Zhejiang University (*Engineering Science*), 2005. 39(10).

[23] Chan, E. and K.M. Yu, A concurrency control model for PDM systems. *Computers in Industry,* 2007. 58(8-9): p. 823-831.

[24] Meng, X.J., et al. Research on integration platform based on PDM for networked manufacturing. in IEEM 2007: 2007 *IEEE International Conference on Industrial Engineering and Engineering Management*. 2007.

[25] Hou, J., et al. Integration of the CAD/PDM/ERP system based on collaborative design. in *Proceedings - ISECS International Colloquium on Computing*, Communication, Control, and Management, CCCM 2008. 2008.

[26] Qiao, L. and Y. Zhang, Implementation approach to information integration between PDM and ERP systems. Beijing Hangkong Hangtian Daxue Xuebao/*Journal of Beijing University of Aeronautics and Astronautics,* 2008. 34(5): p. 587-591.

[27] Xie, S.Q., W.Z. Yang, and Y.L. Tu, Towards a generic product modelling framework. *International Journal of Production Research*, 2008. 46(8): p. 2229-2254.

[28] Xie, S.Q., X. Xu, and Y.L. Tu, A reconfigurable platform in support of one-of-a-kind product development. *International Journal of Production Research*, 2005. 43(9): p. 1889-1910.

[29] Industrial Automation Systems and Integration: Product Data Representation and Exchange: *Part 11: Description Methods*: *The EXPRESS Language Reference Manual*, Reference Number: ISO 10303-11:1994(E), 1994.

[30] Kahn, H., et al., A generic framework for transforming EXPRESS information models. *CAD Computer Aided Design*, 2001. 33(7): p. 501-510.

[31] Industrial automation systems and integration: Product data representation and exchange: *Part 41: Integrated generic resource: Fundamentals of product description and support,* Reference number: ISO 10303-41:2000(E), 2000.

[32] Industrial Automation Systems and Integration: Product Data Representation and Exchange: *Part 45: Integrated Generic Resource: Materials*, Reference Number: ISO 10303-45:1998(E), 1998.

[33] Industrial automation systems and integration: Product data representation and exchange: *Part 203: Application protocol: Configuration controlled 3D designs of mechanical parts and assemblies*, Reference number: ISO 10303-203:1994(E), 1994.

[34] Industrial Automation Systems and Integration: Product Data Representation and Exchange: *Part 49: Integrated generic resources: Process structure and properties*, Reference number: ISO 10303-49:1998(E), First edition, Switzerland., 1998.

[35] Industrial automation systems and integration: Product data representation and exchange: *Part 238: Application protocol:Application interpreted model for computerized numberical controlers*. Reference number:ISO 10303-238, 2007.

[36] Industrial automation systems and integration: Product data representation and exchange: *Part 28: Implementation methods: XML representations of EXPRESS schemas and data*, Reference number: ISO 10303-28:2003(E), 2003.

INDEX

D

E

F

G

H

I

J

L

M

Q

R

S

T

U

V